Marine Aquaculture

Opportunities for Growth

Committee on Assessment of Technology and Opportunities for
Marine Aquaculture in the United States

Marine Board
Commission on Engineering and Technical Systems
National Research Council

NATIONAL ACADEMY PRESS
Washington, D.C. 1992

NATIONAL ACADEMY PRESS • 2101 Constitution Avenue, N.W. • Washington, D.C. 20418

NOTICE: The project that is the subject of this report was approved by the Governing Board of the National Research Council, whose members are drawn from the councils of the National Academy of Sciences, the National Academy of Engineering, and the Institute of Medicine. The members of the panel responsible for the report were chosen for their special competencies and with regard for appropriate balance.

This report has been reviewed by a group other than the authors according to procedures approved by a Report Review Committee consisting of members of the National Academy of Sciences, the National Academy of Engineering, and the Institute of Medicine.

The National Academy of Sciences is a private, nonprofit, self-perpetuating society of distinguished scholars engaged in scientific and engineering research, dedicated to the furtherance of science and technology and to their use for the general welfare. Upon the authority of the charter granted to it by the Congress in 1863, the Academy has a mandate that requires it to advise the federal government on scientific and technical matters. Dr. Frank Press is president of the National Academy of Sciences.

The National Academy of Engineering was established in 1964, under the charter of the National Academy of Sciences, as a parallel organization of outstanding engineers. It is autonomous in its administration and in the selection of its members, sharing with the National Academy of Sciences the responsibility for advising the federal government. The National Academy of Engineering also sponsors engineering programs aimed at meeting national needs, encourages education and research, and recognizes the superior achievements of engineers. Dr. Robert M. White is president of the National Academy of Engineering.

The Institute of Medicine was established in 1970 by the National Academy of Sciences to secure the services of eminent members of appropriate professions in the examination of policy matters pertaining to the health of the public. The Institute acts under the responsibility given to the National Academy of Sciences by its congressional charter to be an adviser to the federal government and, upon its own initiative, to identify issues of medical care, research, and education. Dr. Kenneth I. Shine is president of the Institute of Medicine.

The National Research Council was organized by the National Academy of Sciences in 1916 to associate the broad community of science and technology with the Academy's purposes of furthering knowledge and advising the federal government. Functioning in accordance with general policies determined by the Academy, the Council has become the principal operating agency of both the National Academy of Sciences and the National Academy of Engineering in providing services to the government, the public, and the scientific and engineering communities. The Council is administered jointly by both Academies and the Institute of Medicine. Dr. Frank Press and Dr. Robert M. White are chairman and vice-chairman, respectively, of the National Research Council.

The program described in this report was supported by the U.S. Department of Agriculture and the National Oceanic and Atmospheric Administration under Cooperative Agreement No. 14-35-0001-30475 between the Minerals Management Service of the U.S. Department of the Interior and the National Academy of Sciences.

Library of Congress Cataloging-in-Publication Data

National Research Council (U.S.). Committee on Assessment of Technology and Opportunities for Marine Aquaculture in the United States.

Marine aquaculture: opportunities for growth: report of the Committee on Assessment of Technology and Opportunities for Marine Aquaculture in the United States, Marine Board, Commission on Engineering and Technical Systems, National Research Council.

p. cm.

Includes bibliographical references.

ISBN 0-309-04675-0 : $24.95

1. Mariculture—United States. 2. Mariculture—Government policy—United States. I. Title.

SH138.N38 1992

338.3'71'0973—dc20 92-7308

CIP

This book is printed with soy ink on acid-free recycled stock.

S-512

Printed in the United States of America

COMMITTEE ON ASSESSMENT OF TECHNOLOGY AND OPPORTUNITIES FOR MARINE AQUACULTURE IN THE UNITED STATES

ROBERT B. FRIDLEY, NAE, *Chairman,* University of California, Davis
JAMES L. ANDERSON, University of Rhode Island, Kingston
JORDAN N. BRADFORD, Bradford Seafood, Inc., Pass Christian, Mississippi
BILIANA CICIN-SAIN, University of Delaware, Newark
PAMELA HARDT-ENGLISH, Pharmaceutical and Food Specialists, San Jose, California
BILL L. HARRIOTT, R&D Consultant, Las Cruces, New Mexico (to March 1990)
G. JOAN HOLT, The University of Texas at Austin, Port Aransas
JAMES E. LANNAN, JR., Oregon State University, Newport (to August 1990)
RONALD D. MAYO, J. M. Montgomery, Consulting Engineers, Bellevue, Washington
PETER G. PIERCE, Ocean Products, Inc., Portland, Maine (to January 1990)
KENNETH J. ROBERTS, Louisiana State University, Baton Rouge
JOHN H. RYTHER, Woods Hole Oceanographic Institution (ret.), Woods Hole, Massachusetts
PAUL A. SANDIFER, South Carolina Wildlife and Marine Resources Department, Charleston
EVELYN S. SAWYER, Sea Run Holdings, Inc., Kennebunkport, Maine
R. ONEAL SMITHERMAN, Auburn University, Alabama (to March 1990)

Liaison Representatives

MERYL BROUSSARD, U.S. Department of Agriculture (since July 1991)
BEN DRUCKER, National Marine Fisheries Service, NOAA (to September 1990)
JAMES P. McVEY, Sea Grant College Program, NOAA (to December 1990)
JAMES MEEHAN, National Marine Fisheries Service, NOAA (since September 1990)
JOAN R. MITCHELL, National Science Foundation
JOHN G. NICKUM, U.S. Fish and Wildlife Service (from January 1991)
R. ONEAL SMITHERMAN, U.S. Department of Agriculture (to July 1991)
ROBERT E. STEVENS, U.S. Fish and Wildlife Service

Staff

Susan Garbini, Staff Officer (since July 1990)
Paul M. Scholz (Staff Officer to August 1990), Consultant
Delphine D. Glaze, Project Assistant

MARINE BOARD

Preface

BACKGROUND

Marine aquaculture—the farming of marine finfish, shellfish, crustaceans, and seaweed, as well as ocean ranching of anadromous fish—is a rapidly growing industry in many parts of the world. In some countries, such as Norway and Japan, that have invested in technology development and applications, the marine aquaculture industries represent a substantial sector of the economy. In many developing countries, aquaculture plays an important role in rural development projects and as a commercial enterprise for export markets.

The culture of marine organisms is projected to increase as new technologies are developed that improve the economic feasibility of these operations, as better understanding of the biology and ecology of target species is obtained, and as harvests of wild fish stocks level off or decline. Demand for fish, shellfish, and marine plant products is increasing rapidly in the United States. Domestic per capita consumption is anticipated to continue to grow at about 3 percent per year. Alternatives for meeting increasing demand include more imports, development of nontraditional fishery resources, and expansion of domestic marine aquaculture operations. Another area of significant potential for marine and freshwater aquaculture is an expanded public role in mitigation for loss of habitat or for restoring threatened or overfished wild stocks.

SCOPE OF THE STUDY

The National Research Council (NRC) convened a committee under its Marine Board to assess technology and opportunities for marine aquaculture in the United States. Biographies of committee members appear in Appendix E. The committee was asked to define the national interest in marine aquaculture; to assess the state of practice; and to identify opportunities, establish requirements, and recommend strategies for the appropriate advancement of marine aquaculture in the United States.

The membership of the committee included expertise in aquacultural engineering, aquacultural production, civil engineering, sanitary engineering, fisheries biology, fisheries management, economics, and ocean and coastal policy. Three individuals from private sector marine aquaculture operations served on the committee. Care was taken to ensure a balance of experience in different regions, with different species, and with different aquaculture technologies. The committee was assisted by liaison representatives from federal agencies with related programs or missions: the U.S. Department of Agriculture, the National Oceanic and Atmospheric Administration, the U.S. Fish and Wildlife Service, and the National Science Foundation. The principle guiding the committee, consistent with NRC policy, was not to exclude any information, however biased, that might accompany input vital to the study, but to seek balance and fair treatment.

The primary objective of the study was to identify and appraise opportunities for technology development that can optimize cost-effectiveness and productivity, mitigate environmental constraints, or resolve institutional and policy issues that present obstacles to the advancement of marine aquaculture in the United States. Such an approach does not imply that all problems are susceptible to technological solutions. It seeks only to identify those that might be and to describe possible technological solutions. The committee reviewed national, state, and local policies that regulate or otherwise affect marine aquaculture to determine changes that might be appropriate.

STUDY METHOD

The committee obtained information for the assessment through several approaches. First, it held regional meetings at which members heard formal presentations by practitioners of marine aquaculture business and research activities and those with policy or management oversight. In these workshops, the engineering state of practice of various regional and species-specific marine aquaculture systems was described, along with the economic and institutional factors. These investigations consisted of presentations by invited guests from the West Coast, the Northeast, the South Atlantic, and the Gulf regions on issues specific to these areas. Participants included representatives from industry and government, as well as from institutions conducting relevant research in each region. A second approach to gathering information was through participation of committee members at national meetings of aquaculture organizations. Working groups were convened by the committee, and individuals with specific expertise were invited from around the world to focus on specific technologies and issues of importance to the study. Participants in these sessions are listed in Appendix F.

Individual committee members prepared substantial review papers on all the major areas of information addressed in the study. A bibliography of reference material used in these preliminary papers is included at the end of this report. Additional information on specific topics was solicited by the committee from environmental and conservation organizations, and organizations representing traditional fisheries industries. Representatives of the major government agencies with responsibilities for oversight of marine aquaculture also participated in the committee's deliberations.

ORGANIZATION OF THE REPORT

The final report represents a synthesis of information gathered by the committee, aimed at focusing this wide-ranging material into an examination of the present status of marine aquaculture in the United States and the major obstacles to its emergence as a successful industry. The report is not intended as a comprehensive survey, although overviews of world (Appendix A) and U.S. (Appendix B) aquaculture are presented in the appendixes. Appendix C is an authored paper examining the sociocultural aspects of U.S. marine aquaculture.

Chapter 1 introduces the major issues to be addressed in the report. Chapter 2 reviews the status of world and U.S. aquaculture, with an emphasis on marine aquaculture and comparisons between world and U.S. production and with a focus on the economic contribution. Chapter 3 presents an overview of the federal and state policy framework in which marine aquaculture has operated over the past 15 years, and addresses continuing and newly emerging problems that constrain the growth of this industry in the United States, including conflicts among various users, coastal management issues, and the role of state and local governments. Environmental issues are examined separately in Chapter 4. Chapters 5 and 6 provide a detailed examination of the scientific, technical, and educational base that is needed to build a successful marine aquaculture industry, both for resolving environmental problems and for achieving economic feasibility. The major conclusions and recommendations that follow from the findings of this investigation are presented in Chapter 7. An Executive Summary provides a synopsis of the report. An extensive bibliography of reference material on the subject matter is also included.

This report is intended to serve as a guide to federal and state government agencies and the private sector in making decisions about appropriate policy, regulatory, and economic actions that are needed to improve the prospects for success of the U.S. marine aquaculture industry. It is hoped that the report will also serve as an educational document for the public, the media, and those who are involved in any aspects of marine aquaculture.

ACKNOWLEDGMENTS

The committee appreciates greatly the valuable input and insight that was provided by participants in two of its regional meetings and in two committee meetings that were convened at gatherings of international technical societies. Participants in these sessions are listed in Appendix F.

The committee chair wishes to extend a particular thanks to Professor Raul Piedrahita of the Agricultural Engineering Department at the University of California, Davis for significant and valued input, particularly to Chapter 5. Special thanks are due to Lucy Garcia, principal staff assistant for the Aquaculture and Fisheries Program, University of California, Davis, for exceptional support and liaison activities on behalf of the committee chair during the conduct of this study. The committee also wishes to thank Jean-Pierre Plé, doctoral student in the College of Marine Studies at the University of Delaware, for his assistance in the preparation of Appendix A (Review of World Aquaculture) and Appendix C (Federal Marine Aquaculture Policy). The rewards for committee service are not great; however, the rewards of working with and getting to know a group of knowledgable and dedicated people who are willing to voluntarily support and provide input to such a committee are very much worthwhile.

Finally, the committee wishes to acknowledge the valuable input of reviewers who did an exceptionally good job of providing specific suggestions and corrections, of the liaisons from federal agencies who provided extensive information about the public sector's efforts, and of Marine Board staff, whose support was invaluable.

Contents

Marine Aquaculture

Executive Summary

Marine aquaculture, the farming of marine finfish, shellfish, crustaceans, and seaweed, as well as the ocean ranching of anadromous[1] fish, is a rapidly growing industry in many parts of the world. In the United States, freshwater aquaculture (primarily the farming of catfish, trout, crayfish, and ornamental fish) is an expanding industry; however, marine aquaculture has yet to achieve economic success beyond a limited basis. Constraints to the industry have included difficulties and costs of using coastal and ocean space, public concerns about environmental effects of wastes on water quality, conflicts with other users of the coastal zone (e.g., boaters and fishermen), objections to marine aquaculture installations on aesthetic grounds from coastal property owners, and broad ecological issues involving concerns about genetic dilution of wild stocks and transfer of diseases by cultured species through escapement of cultured animals. Poor water quality, high labor and land costs, and limited warm water temperatures also inhibit the success of marine aquaculture in the United States.

On the other hand, the consumption of seafood in the United States is increasing at the same time that yields from capture fishing are reaching the limits of sustainable returns, and the nation relies increasingly on imports to meet the growing consumer demand for seafood. The opportunity, therefore, exists for U.S. aquaculture to develop the capability to supply this growing demand and for marine aquaculture to make a significant contribution.

The National Research Council convened a committee under its Marine Board to assess the technology and opportunities for marine aquaculture in the United States. The primary objective of the study was to identify and appraise opportunities for technology development that can optimize the cost-effectiveness and productivity of marine aquaculture in the United States, as well as engineering and policy actions that would address associated environmental concerns.

The committee concluded that a number of benefits will accrue to the nation from the addition of an economically viable, technologically advanced, and environmentally sensitive healthy marine aquaculture industry. These benefits include providing wholesome seafood to replace declining harvests of wild fish, products for export to improve the nation's balance of trade, enhancement of commercial and recreational fisheries and fisheries that are overfished or otherwise threatened, economic opportunities for rural communities, and new jobs for skilled workers, particularly in coastal communities where some traditional fisheries are at maximum sustainable yield or in decline. The advancement of the science and technology base in marine aquaculture also will provide benefits to other industries, such as biotechnology and pharmaceuticals.

The prospects for marine aquaculture as an emerging enterprise are uncertain and depend on whether a number of problems are resolved. However, given a fair share of support for the development of an advanced scientific and engineering base, as well as a reasonable and predictable regulatory framework, many of the problems that presently constrain marine aquaculture could be resolved.

Although legislation to promote aquaculture was passed in 1980 (National Aquaculture Act, P.L. 96-362) and again in 1985 (National Aquaculture Improvement Act, P.L. 99-198), a number of problems have prevented these expressions of policy intent from effectively transforming marine aquaculture into a dynamic industry. First, no funds were ever appropriated to agencies to implement the provisions of these acts. Second, the needs of marine aquaculture have tended to be overshadowed by the interests of the freshwater aquaculture industry, which are more closely linked to those of the traditional agriculture community through its geographic focus in inland farming areas. Moreover, marine aquaculture, because of its location in the coastal zone, operates under a complex coastal regulatory regime, and tends to arouse intense scrutiny because of widespread public concern about activities that take place in or near the ocean.

CONCLUSIONS

The present study investigated the opportunities for improving the outlook for U.S. marine aquaculture and concluded that the issues that constrain its development will need to be specifically addressed through three primary avenues: (1) advances in the scientific, technical, and engineering base that underlies this industry, both to achieve more cost-effective operations and to mitigate environmental problems; (2) changes in federal and state agency roles to provide a regulatory and funding framework that encourages the industry's growth while ensuring that environmental concerns are addressed; and (3) congressional actions to attend to a number of unresolved policy issues. Achieving these objectives will depend on active congressional oversight of the executive agencies charged with implementing the national policies expressed in the National Aquaculture Act and the National Aquaculture Improvement Act.

RECOMMENDATIONS

Advances in Technology and Engineering— A Marine Aquaculture Initiative

The opportunity exists for technology and increased knowledge to provide solutions to many of the environmental, economic, and biological limitations that constrain marine aquaculture's transformation into a significant U.S. industry. The design of new technologies can play a key role in improving all aspects of culture operations and auxiliary systems that will contribute to the economic feasibility of marine aquaculture, as well as alleviate environmental problems. The opportunity, however, can be realized only if federal policy and action strongly support the development of needed technology.

The committee recommends that Congress make a $12 million national commitment to a strategic R&D initiative that will support the research necessary to develop marine aquaculture technology, to address environmental issues and concerns and to provide economical systems. Leadership in this initiative should be provided by the U.S. Department of Agriculture (USDA) with coordination by the Joint Subcommittee on Aquaculture (JSA), and implemented under memoranda of understanding among federal agencies that sponsor or conduct research related to marine aquaculture. The initiative should address the following research and development needs:

- the interdisciplinary development of environmentally sensitive, sustainable systems that will enable significant commercialization of onshore (on land) and nearshore marine aquaculture without unduly increasing conflict over the use of coastal areas;
- development of the biological and engineering knowledge base for technologies and candidate species needed to make decisions regarding commercialization of offshore marine aquaculture operations that avoid the environmental impacts of nearshore operations;
- creation of (1) technology centers to be used for these technology development programs, and (2) marine aquaculture parks for fostering new environmentally sensitive commercial development;

• design and implementation of improved higher education programs;

• new and improved procedures and systems to collect and exchange data and technical information; and

• promotion of marine aquaculture as a vital component of fisheries stock mitigation and enhancement by (1) facilitating aquaculture's role in the preservation of threatened or endangered species populations and of genetic diversity, including greater involvement of private sector facilities; (2) developing production procedures for the broader range of species necessary for effective mitigation of negative impacts on fish and shellfish stocks; and (3) developing and implementing improved methods for determining the effectiveness of using cultured stock for fish and shellfish enhancement activities in support of commercial, recreational, and ecological purposes.

Federal Agency Actions

The federal agencies with primary jurisdiction over marine aquaculture activities include the U.S. Department of Agriculture (USDA), the U.S. Fish and Wildlife Service (FWS), and two branches of the National Oceanic and Atmospheric Administration (NOAA)—the National Marine Fisheries Service (NMFS) and the National Sea Grant College Program. Although, the USDA was designated as lead agency in the National Aquaculture Improvement Act of 1985, it is unrealistic to expect that the FWS and NOAA will give up their long-standing interests in this domain; however, more effective means of coordinating their activities need to be developed. More active leadership and more effective coordination of federal activities under congressional oversight are necessary to translate the intent of existing national legislation regarding aquaculture into positive actions.

U.S. Department of Agriculture

• It is recommended that the lead role of the USDA be strengthened by establishing a formal entity focused on aquaculture, at an appropriately high level in the agency, and by acquiring expertise in marine aquaculture throughout USDA's services. Specific additional funds need to be allocated to target marine aquaculture activities in existing USDA services.

• It is recommended that the USDA be charged with leadership in the promotion of commercial aquaculture including the research and support services (i.e., National Aquaculture Information Center) required, particularly in the areas of production, processing, distribution, and marketing of marine aquaculture products, especially as food products.

• It is recommended that, under the leadership of the USDA, several interagency memoranda of understanding (MOUs) be created to clarify the missions, roles, and responsibilities of each agency with respect to aquaculture and specifically marine aquaculture.

Joint Subcommittee on Aquaculture

It is recommended that in addition to its current role as a forum for interagency discussion, the JSA be charged with designing a streamlined planning and permitting process for marine aquaculture activities emphasizing joint local, state, and federal coordination, and take responsibility for promoting the inclusion of marine aquaculture in the Coastal Zone Management Act.

The JSA should also conduct a comprehensive evaluation of impacts of the Lacey Act (P.L. 97-79, as amended in 1981) on marine aquaculture, and make recommendations to Congress for appropriate changes to specifically encourage development of marine aquaculture based on ecologically sound considerations.

U.S. Fish and Wildlife Service

It is recommended that the FWS continue to exercise leadership in the area of fisheries enhancement of anadromous species. Such leadership should include:

- promoting the use of private aquaculture for enhancement of stocks of various anadromous species that are heavily fished or otherwise threatened or endangered;
- supporting the development of technology for rearing and releasing anadromous stocks where needed; and
- administering the introduction and transfer of nonindigenous anadromous species.

*National Oceanic and Atmospheric Administration/
National Marine Fisheries Service*

It is recommended that NOAA/NMFS be charged with leadership in the management and assessment of stock-enhanced marine fisheries. Such leadership should include:

- evaluating the effectiveness of existing and future stock enhancement programs;
- supporting the development of technology for (1) producing juvenile stocks needed for nonanadromous marine fisheries enhancement and related aquaculture, and (2) releasing marine stocks, where needed;

• assessing the impact (or potential impact) of various nearshore and offshore marine aquaculture practices on the marine environment and fisheries; and

• administering the introduction and transfer of nonindigenous marine species.

National Oceanic and Atmospheric Administration/ National Sea Grant College Program

It is recommended that NOAA/Sea Grant be charged with leadership in support of research and extension programs on marine aquaculture-related topics focused on preservation of the marine environment, understanding the life history of candidate species, and multiple use of marine resources, including associated social, economic, and policy issues. Candidate research topics include:

• environmentally safe technology, methods, and systems for culturing marine species in the marine environment;

• marine aquaculture technology that is synergistic with other uses of the sea (i.e., multiple use technologies);

• life history and developmental biology of candidate species;

• the socioeconomic dynamics of the marine aquaculture industry (e.g., effects on local employment patterns);

• methods for addressing and resolving conflicts between marine aquaculture and other competing users of the marine environment;

• comparative studies of state practices regarding the regulation and promotion of marine aquaculture; and

• alternative institutional and policy structures for managing marine aquaculture in other countries.

Congressional Action

The development of marine aquaculture is beset with complexity that stems from unique factors that distinguish it from other kinds of agricultural activity. These are:

• the interaction of marine aquaculture with other marine and coastal activities and interests—interactions often characterized by conflict;

• the fact that although marine aquaculture is ocean based, it depends on the use of land and freshwater resources as well; and

• the numerous environmental and regulatory considerations involved in the development and use of coastal zone land and water.

This complexity entails the involvement of a number of federal, state, and local agencies that are responsible for all aspects of the advocacy, promotion, conduct, and regulation of marine aquaculture, leading to an array of planning acts, policies, and regulations. For marine aquaculture to realize its potential, it needs to be addressed explicitly within a coordinated and coherent policy framework in federal, regional, and state ocean and coastal zone planning activities.

Although most of the recommendations outlined above can be implemented by the designated agencies through MOUs and by the JSA under existing legislation, three unresolved policy issues need to be addressed by Congress.

Completion of the Federal Policy Framework for Marine Aquaculture

Coastal Zone Marine aquaculture must be explicitly included in coastal zone plans that ensure its proper consideration and evaluation in development and environmental decisions. It is recommended that Congress designate marine aquaculture as a recognized use of the coastal zone in the Coastal Zone Management Act (P.L. 92-583, as amended in 1990, P.L. 101-508). Such designation will stimulate states to include marine aquaculture in state coastal management plans for achieving a balanced approach to land use, resource development, and environmental regulation.

Federal Waters Currently, no formal framework exists to govern the leasing and development of private commercial aquaculture activities in public waters. A predictable and orderly process for ensuring a fair return to the operator and to the public for the use of public resources is necessary to the development of marine aquaculture. It is recommended that Congress create a legal framework to foster appropriate development, to anticipate potential conflicts over proposed uses, to assess potential environmental impacts of marine aquaculture, to develop appropriate mitigation measures for unavoidable impacts, and to assign fair public and private rents and returns on such operations.

Revision of Laws That Impede the Development of Marine Aquaculture

The Lacey Act Environmental preservation and the protection of indigenous species are important concerns; however, the Lacey Act (P.L. 97-79, as amended 1981, P.L. 97-79) as presently constituted, creates a barrier to the development of marine aquaculture. Control points for regulation of the movement of living fish between states need to be based on scientific and ecological information rather than solely on state borders. It is recommended that the Joint Subcommittee on Aquaculture conduct a comprehensive review of the Lacey Act to recommend to Congress revisions that could encourage the development of marine aquaculture within an environmentally sound regulatory framework.

Creation of a Congressional Committee or Subcommittee on Aquaculture

As human demand for seafood exceeds sustainable yield from traditional fisheries, dependence on capture fisheries is likely to shift to dependence on aquaculture. No formal mechanism currently exists for congressional policymakers to anticipate this transition and make appropriate policy decisions; nor is there a mechanism for congressional oversight of the federal agency and JSA actions mandated by the National Aquaculture Act and its amendments. It is recommended that Congress consider creating an oversight committee or subcommittee on aquaculture to provide a formal linkage between the House Agriculture Committee and the House Merchant Marine and Fisheries Committee to ensure the implementation of existing and future policies enacted to promote aquaculture.

CONCLUSION

A number of benefits will accrue to the nation from the addition of an economically vital, technologically advanced, and environmentally sensitive marine aquaculture industry. The prospects of this emerging enterprise are for healthy and vigorous growth, given a fair share of support for the development of an advanced scientific and engineering base, along with a reasonable and predictable regulatory framework. On this basis, the environmental problems that presently constrain marine aquaculture are likely to be resolved so that it can contribute to the continued vitality of the nation's living marine resources.

NOTE

1. Fish that ascend rivers from the sea at certain seasons for breeding (e.g., salmon and shad).

1

Introduction

Aquaculture, the husbandry of aquatic animals and plants, has been practiced since the earliest records of human history and is a rapidly growing industry in many parts of the world. In Norway and Japan, for example, the marine aquaculture industries comprise a significant economic sector in the national economies. Freshwater culture of fish and crustaceans (primarily catfish and crayfish) represents the fastest growing agricultural industry in the United States (DeVoe and Mount, 1989; Gulf States Marine Fisheries Commission, 1990).

The National Aquaculture Act (P.L. 96-362), signed into law in 1980, states that national policy is "to encourage the development of aquaculture in the United States." This initiative was primarily a response to (1) a growing concern that natural harvests of fisheries would shortly reach their maximum sustainable yields and (2) the steadily increasing negative annual trade deficits in fish and fish products. Despite this legislation, a review of the nation's aquaculture industry in 1983 under the auspices of the Joint Subcommittee on Aquaculture (JSA) (National Aquaculture Development Plan, 1983) noted that many impediments to the expansion of aquaculture that had been identified in an earlier National Research Council (NRC) report (NAS, 1978) still persisted. In the nearly 15 years that now have elapsed since publication of the NRC report, progress toward the establishment of a successful industry remains slow.

In general, the U.S. aquaculture industry has failed to capture a significant share of the potential domestic or global market. In particular, the U.S. marine aquaculture industry—the farming or ranching of marine finfish and shellfish—has yet to demonstrate long-term economic viability.

The following chapters are the products of an investigation to ascertain the present state of practice of marine aquaculture technology; to identify and appraise technical, social, environmental, and institutional issues constraining the advance of this industry; and to recommend technological and policy strategies that might lead to improved prospects in the future.

WORLD/U.S. AQUACULTURE PRODUCTION

Although farming of aquatic animals and plants is a practice equally as old as farming on land, as a modern industry, aquaculture is still a relatively minor source of food and other products compared to agriculture and traditional capture fisheries. Total world fish production in 1988 was reported by the Food and Agriculture Organization of the United Nations (FAO) at 98 million metric tons (mmt), of which 14 mmt were from aquaculture. The economic value of the 1988 world aquaculture crop of 14 mmt is estimated to be $22.5 billion, an increase of 19 percent from the $18.8 billion value of the 1987 crop (FAO, 1990).

Of the roughly 300,000 metric tons of aquatic life grown for food in the United States in 1988, three-quarters or more were freshwater organisms. U.S. marine aquaculture production in 1988 was about 75,000 metric tons, of which approximately 80 percent were oysters. All of the marine species, except oysters, are in the early stages of commercial development in the United States, and most projects have yet to achieve sustained economic viability. U.S. marine aquaculture has not expanded in accordance with the growth of the world industry during recent history. In fact, in some areas, the U.S. market share has declined (e.g., cultured salmon and oyster production).

THE NATIONAL INTEREST IN MARINE AQUACULTURE

During the 1980s, per capita consumption of seafood increased steadily until 1987 then stabilized (Figure 1-1 and Table 1-1). However, when considering a more significant indication of demand—per capita expenditure on seafood (Figure 1-2)—one finds a steadily upward trend throughout the decade (except for 1988). Although the growth rate of seafood expenditure has slowed, U.S. consumers are still spending more for seafood despite the recession, recent highly publicized seafood safety issues, and the fact that prices are increasing faster than general consumer prices (Table 1-1). Many observers expect that seafood expenditures will continue to grow.

At the same time that marine aquaculture has lagged, consumer demand for seafood has increased concomitantly with population growth and on a per capita basis as well. Concerns about improved nutrition and lowered cholesterol intake have stimulated changing dietary habits that include in-

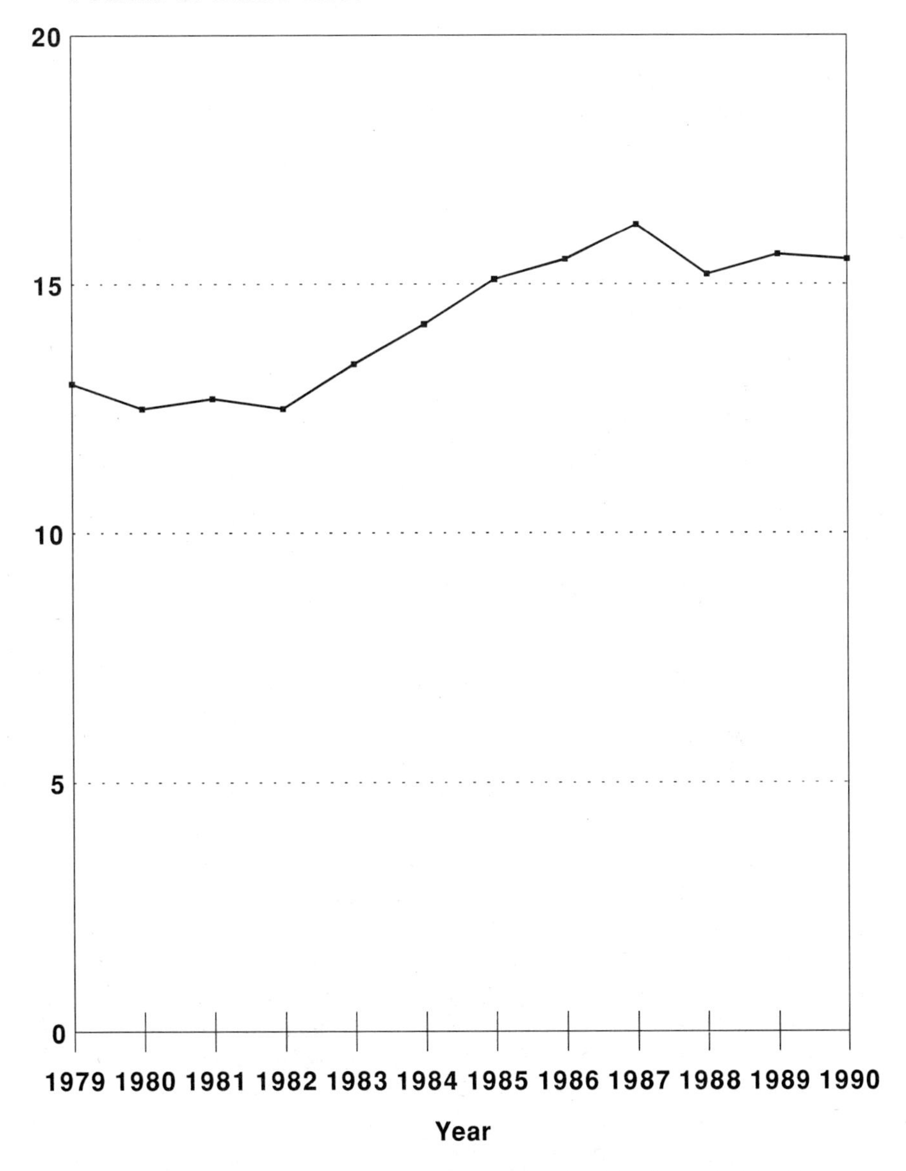

FIGURE 1-1 U.S. per capita consumption of seafood 1979–1990. SOURCE: U.S. Department of Commerce (1990).

TABLE 1-1 U.S. Consumption of Seafood 1979–1990: Per Capita Seafood Consumption, Seafood Prices, and Derived Per Capita Seafood Expenditures

Year	U.S. Per Capita Seafood Consumption (lbs edible meat)[a]	U.S. Per Capita Seafood Consumption Index (1982–1984=100)[b]	U.S. Consumer Price Index, Fish (1982–1984=100)[c]	Derived U.S. Per Capita Seafood Expenditure Index (1982–1984=100)[d]	U.S. Consumer Price Index, All Food (1982–1984=100)[c]
1979	13.0	97.2	80.1	77.9	79.9
1980	12.5	93.5	87.5	81.8	86.8
1981	12.7	95.0	94.8	90.0	93.6
1982	12.5	93.5	98.2	91.8	97.4
1983	13.4	100.2	99.3	99.5	99.4
1984	14.2	106.2	102.5	108.9	103.2
1985	15.1	112.9	107.5	121.4	105.6
1986	15.5	115.9	117.4	136.1	109.0
1987	16.2	121.2	129.9	157.4	113.5
1988	15.2	113.7	137.4	156.2	118.2
1989	15.6	116.7	143.6	167.6	125.1
1990	15.5	115.9	146.7	170.1	132.4
Change 1980–1990	+24%		+67.7%	+107.9%	+52.5%

[a]U.S. Department of Commerce (1990).

[b]Calculated from U.S. per capita seafood consumption reported by U.S. Department of Commerce (1990).

[c]Putnam and Allshouse (1991).

[d]U.S. Department of Commerce (1990); Putnam and Allshouse (1991). Derived U.S. per capita seafood expenditure index calculated as (U.S. per capita seafood consumption index × U.S. consumer price index for fish)/100.

creased consumption of seafood. In addition, immigration of large numbers of cultural groups from East and Southeast Asia has created a growing consumer market for seafood. The increased consumer demand comes at a time when yields from capture fishing are beginning to peak, an opinion reinforced by the consensus that virtually all of the established major world fisheries and most of the recently discovered and exploited resources are already fished at, if not beyond, their limit of sustainable yield (Royce, 1989). At any rate, U.S. capture fisheries have not been able to provide for increased consumer demand. In 1988 the importation of seafood into the United States exceeded 25 percent of the value of all food and live animals imported—seafood importation was a little more than all other animals and meat products ($5.3 billion versus $5.2 billion, respectively) and a little less than all fruits, vegetables, and nuts ($5.4 billion) (U.S. Bureau of the Census, 1988).

The opportunity, therefore, exists for U.S. aquaculture to develop the capability to supply this growing demand, and marine aquaculture represents a significant part of this opportunity. A vital marine aquaculture industry could also contribute to the nation's welfare in other ways. For example, in coastal regions where traditional fishery jobs are in decline, marine aquaculture provides employment opportunities that maintain links to traditional life-styles. Marine aquaculture provides the basis for rejuvenating the seafood processing industry in some areas, including the production, manufacture, and processing of nonfood products from marine culture, such as pharmaceuticals and ornamental fish. The advantages of marine aquaculture over traditional fisheries for local economies include year-round industries and the development of a technically skilled work force (for further discussion of the social and cultural aspects of marine aquaculture on local economies, see Appendix D).

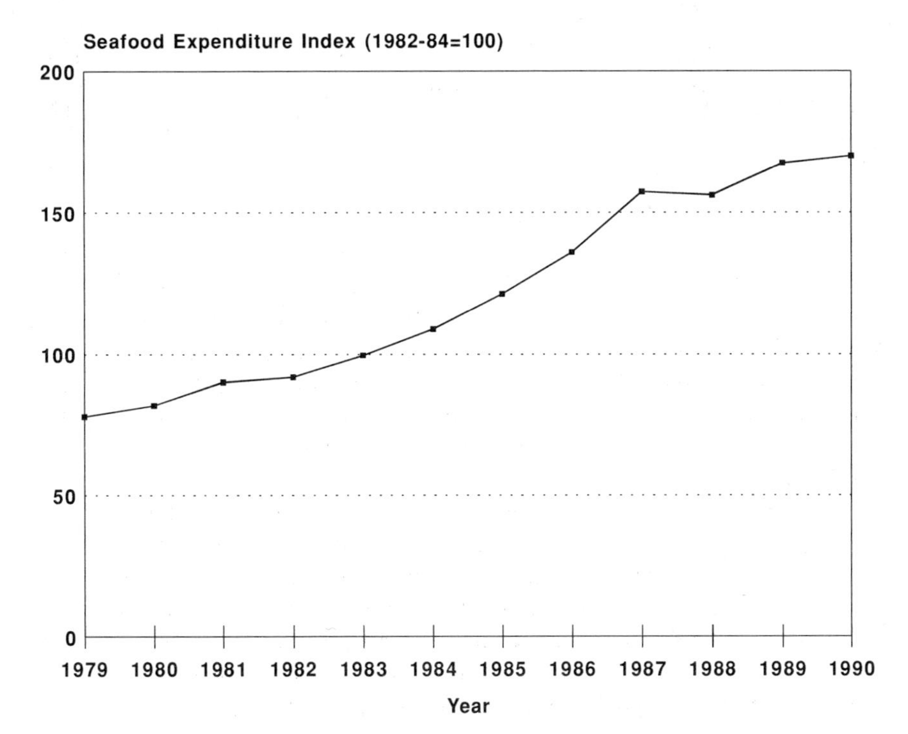

FIGURE 1-2 Derived U.S. per capita seafood expenditure index 1979–1990. Note: Derived U.S. per capita seafood expenditure index calculated as (U.S. per capita seafood consumption index × U.S. consumer price index for fish)/100. SOURCES: U.S. Department of Commerce (1990); Putnam and Allshouse (1991).

Shrimp *(Penaeus vannaemi)* harvested from an intensive culture pond in South Carolina.

For the nation, marine aquaculture has the potential to contribute significantly to the enhancement of fisheries stocks that are in decline or in danger of extinction. This role could be instrumental in augmenting species for recreational purposes, for commercial fisheries, or for wild species preservation. The emergence of fish farming, in concert with wild fisheries that are managed to maintain healthy levels of productivity, could alleviate the conflicts between the need for fish as food and the view of fish as a recreational or aesthetic resource.

As a complex scientific and engineering field, marine aquaculture systems and technology can contribute to the development of marine biotechnology, which in turn has the potential to contribute to a number of medical and scientific advances. The sector of the U.S. marine aquaculture industry that is focused on designing and engineering operating systems is experiencing a growth in demand for export of systems and expertise to other countries seeking to establish technology-based marine aquaculture operations.

PROBLEMS AND CONSTRAINTS

Given the existence of growing consumer demand for the product and the capability of U.S. science and engineering to design and operate advanced

Harvesting red drum *(Sciaenops ocellatus)*.

systems, why has U.S. marine aquaculture lagged behind other countries in productivity and profitability?

This subject was addressed by the NRC nearly 15 years ago in a study by the Board on Agriculture and Renewable Resources on the broader topic of aquaculture (NAS, 1978). A major conclusion from this investigation was that "constraints on orderly development of aquaculture tend to be political and administrative, rather than scientific and technological." Among those identified were multiple-use conflicts, legal constraints, and difficulty in locating capital for entrepreneurial investment.

This report examines the widespread view that despite progress in the area of national policy for aquaculture—passage of the National Aquaculture Act of 1980 (P.L. 96-362), assignment of a lead role in encouraging the

Culture of ornamental reef fish—pygmy angels—in spawning tank.

Net cage culture of Atlantic salmon in Norway.

industry to the Department of Agriculture (USDA), establishment of an interagency Joint Subcommittee on Aquaculture (JSA), and the preparation of a National Aquaculture Development Plan (NADP) in 1983—a number of problems continue to prevent the successful growth of this fledgling industry in the United States.

As noted in previous studies, many of the problems tend to be in the institutional and legal realm and need to be addressed through federal agency leadership. For example, a number of local, state, and federal agencies are involved in granting permits and licenses for marine aquaculture activities. The process could be streamlined through the establishment of federal model guidelines and procedures. Prohibitions on interstate transport of aquaculture products and other interstate issues may have to be reexamined at the federal level based on current scientific understanding of actual risks. From the viewpoint of many experts on coastal resources policy, marine aquaculture needs to be included in a national framework for managing coastal resources that balances competing uses and values in the national interest in order to provide a level of predictability necessary for planning commercial aquaculture ventures.

At the state and local levels, the issue of property rights for the marine aquaculture industry, including the leasing of submerged lands and/or the water column, remains unaddressed in most states and is a major disincentive to would-be entrepreneurs who have no legal means of protecting the products of their endeavors. The states are also involved in the resolution of conflicts among competing users of coastal areas, such as capture fisheries and recreational interests as well as marine aquaculture operations.

Land and water use conflicts are major constraints to the development of the marine aquaculture industry. They arise from multiple-use conflicts with commercial and recreational fishing and also from environmental concerns about pollution. Of particular public concern is water pollution from the wastes produced in aquaculture systems, from excess feed that may contain the antibiotics and pesticides used to prevent disease and predation, or from hormones used to stimulate growth.

Other environmental and aesthetic issues also are a serious impediment to the development of marine aquaculture. There is growing controversy about the privatization of public resources (public waters) that occurs in many aquaculture operations such as with salmon net pens, as well as aesthetic objections to these installations. Escapement or release of cultured animals, either accidentally from cages or purposefully, as in ocean ranching, raises fears of genetic dilution of native stocks that might lead to their extinction, the spread of disease from cultured animals to native stocks, or the release of exotic species with possible adverse ecological consequences.

Within the national institutional framework of competing organizations, marine aquaculture is hampered by its relative lack of organizational and political power when compared to the traditional capture fisheries industries, in terms of both a structured marketing network and an effective organizational base from which to promote its interests. The public support system for research and technology development through university research and extension services that accompanied the success of U.S. agriculture also is lacking for marine aquaculture.

Economic factors hamper the success of marine aquaculture, too. For U.S. products to become competitive in the world market, technology and engineering systems will need to be developed to compensate for the relatively high costs of labor and coastal land relative to these costs in other countries where marine aquaculture is more successful. For example, one possible route to improved competitiveness is to develop new species for culture, such as halibut, red drum, red snapper, striped bass, or scallops. These endeavors will depend on extensive research and the development of new and complex engineering systems that, in turn, require a strong science and technology base and a highly skilled work force. Existing technical and university programs do not provide adequate education, training, or research and technology development essential to stimulate the growth of this high-technology industry in the United States. Nor do present extension services provided through federal agencies (e.g., USDA and the National Sea Grant College Program) offer the necessary support in training and technology transfer that are required by a newly emerging technology-based industry.

Subsequent chapters of this report indicate that whether marine aquaculture becomes a successful industry in the United States will depend on the extent to which the following three major problem areas are addressed:

1. the establishment of a policy framework for resolving coastal use conflicts and related institutional obstacles (e.g., management of living marine resources, competition of various users in the coastal zone, delineation of appropriate state and federal government roles);
2. the development and application of new technologies to diminish or mitigate harmful environmental impacts and to establish economic feasibility for aquaculture operations; and
3. the development of dependable and predictable domestic and export markets for marine aquaculture products.

Although all of these issues have policy and regulatory dimensions needing resolution, scientific and technological advances can, in many cases, mitigate some of the problems associated with them. This point is particularly true in the case of concerns about harmful environmental impacts from aquaculture operations and conflicts related to competing uses of coastal land and waters. Solutions to these issues, in turn, would improve the profitability of marine aquaculture operations by decreasing costs of compliance with regulations, increasing profit margins due to more efficient operating systems, and possibly, expanding potential markets based on development of new and improved products.

The prognosis for marine aquaculture is uncertain. The potential for successful growth depends on whether a number of problems and constraints are addressed. Advances in science and technology, although crucial to the economic success of U.S. marine aquaculture, cannot by themselves ensure improved prospects for this industry. Policy initiatives at the state and federal levels will also be necessary.

REFERENCES

DeVoe, M.R., and A.S. Mount. 1989. An analysis of ten state aquaculture leasing systems: Issues and strategies. Journal of Shellfish Research 8(1):233-239.

Food and Agriculture Organization of the United Nations (FAO). 1990. FIDI/C:815 Revision 2, as reported in Fish Farming Internat. 17(8):12-13.

Gulf States Marine Fisheries Commission. 1990. Summary of aquaculture programs by state. A report to the Technical Coordinating Committee. March 14. Orange Beach, Ala.

National Academy of Sciences (NAS). 1978. Aquaculture in the United States: Constraints and Opportunities. Washington, D.C.: National Academy Press.

National Aquaculture Development Plan, Vol. I. 1983. The Joint Subcommittee on Aquaculture of the Federal Coordinating Council on Science, Engineering, and Technology. Washington, D.C.: U.S. Government Printing Office.

Putnam, J.J., and J.E. Allshouse. 1991. Food consumption, prices, and expenditures 1968-1989. Statistical Bulletin No. 825. U.S. Department of Agriculture, Economic Research Service, Washington, D.C.

Royce, W.F. 1989. A history of marine fishery management. Aquatic Sci. 1:27-44.

U.S. Bureau of the Census. 1988. Unpublished data. Microfiche Series, Foreign Trade Division, EM575 and IM-175.

U.S. Department of Commerce. 1990. Fisheries of the United States. National Oceanic and Atmospheric Administration, National Marine Fisheries Service, Washington, D.C.

2

Status of Aquaculture

Marine aquaculture in the United States lags behind that in other developed countries such as Japan and Norway. This situation is not the result of any natural disadvantages for the aquaculture of many marine species; development has been constrained by a number of factors, including the regulatory environment, economic opportunities, and availability of research and educational support. This chapter briefly reviews world and U.S. aquaculture production and then focuses on analysis of the status of marine aquaculture in the United States. Details of world aquaculture production are presented in Appendix A. A review of U.S. freshwater aquaculture is provided in Appendix B.

AN OVERVIEW OF AQUACULTURE AND FISHERIES WORLDWIDE

World aquaculture production in 1988 reached 14 million metric tons (mmt) (FAO, 1990),[1] an increase of about 10 percent over the previous year and a mean annual increase of 7 percent from the 6 mmt reported for 1975 by Pillay (1976). The latter represents a doubling each decade; however, part of the increase may be more apparent than real because the number of countries reporting aquaculture statistics to the Food and Agriculture Organization (FAO) increased from 67 to 144 during that same period (FAO, 1990).

Total annual world harvest from capture fishing also increased by about 7 percent annually, from 21 to 40 mmt during the decade 1950–1960, but production then began to slow down (to 69 mmt by 1968) followed by a decade of ups and downs with little net increase. This slowdown was primarily due to the nearly simultaneous failure of three of the world's largest fisheries (North Atlantic herring, Peruvian anchovetta, and South Atlantic pilchard). These fisheries have since wholly or partially recovered, and world production again began to increase by 1978, but at a reduced annual rate of about 2.5 percent over the next decade.

In 1988, total world fish production was reported by FAO at 98 mmt, a figure that includes 14 mmt from aquaculture and 84 mmt from capture fishing. If the 23 mmt used for industrial purposes (i.e., meal, oil) are excluded, 61 mmt from the commercial fishery were used for direct human consumption. Thus, the aquaculture yield of 14 mmt represents 19 percent of the total edible fish production, or 23 percent of the edible fish taken by commercial fishing in 1988.

Several estimates made during the late 1960s and early 1970s placed the potential yield of fish from the sea at or about 100 mmt (Ricker, 1969; Ryther, 1969; Gulland, 1971), a figure that now appears to be generally accepted (Hjul, 1973; Bailey, 1988). As the 100-mmt yield is approached by landing statistics, many feel that yields from capture fisheries are beginning to peak. This opinion is reinforced by the consensus that virtually all of the established major world fisheries and most of the recently discovered and exploited resources (Bering Sea, Falkland Islands, New Zealand, and the Antarctic) are already fished at, if not beyond, their sustainable yield. With the exception of unconventional resources of doubtful economic or human food value (e.g., Antarctic krill, lantern fish), no major unexploited or underutilized fisheries remain in the sea (Royce, 1989).

The human population has roughly doubled since 1950 (2.5 to 5.0 billion), while world fish production has more than quadrupled (21 to 98 mmt). Thus, annual per capita utilization (as food and industrial products) has also more than doubled (18 to 43 pounds per capita). If the increase in consumption were to continue at the same rate to the year 2000, when the human population is expected to reach 6 billion, an annual production of 138 mmt of fish would be needed. It is doubtful that capture fishing, apparently already reaching its natural limit, could continue to meet such a demand. If aquaculture were to continue to grow at the same rate it has over the past decade, it would produce 33 mmt by the year 2000 and could effectively supplement a commercial fishing yield of 100 mmt in meeting the anticipated demand.

A summary of the 1988 aquaculture yield of 14 mmt, derived from data given by FAO, is shown in Table 2-1. Yields are broken down into major categories, both geographically and by species groups. The East Asian countries of China, Japan, the two Koreas, Taiwan, and the Philippines together account for about three-quarters (11 mmt) of the world's aquaculture production, with China alone accounting for nearly one-half the

TABLE 2-1 World Aquaculture Production, 1988 (million metric tons)

Region	Finfish	Crustaceans	Mollusks	Seaweeds	Total
Africa	0.07				0.07
North America	0.30	0.03	0.10		0.40
Latin America	0.04	0.10	0.05	0.02	0.20
Europe	0.50	0.60			1.00
USSR	0.40				0.40
Near East	0.03				0.03
East Asia	5.00	0.30	2.00	3.50	11.00
West Asia	1.00	0.10	0.10	0.08	1.00
Total	7.00	0.50	3.00	4.00	14.00

NOTE: Figures are rounded.
SOURCE: Food and Agriculture Organization (1990).

world's production (7 mmt). The West Asian countries of Indonesia, Vietnam, Thailand, India, and Bangladesh together grow more than 1 mmt, bringing the Asian total to more than 12 mmt, or 84 percent of the worldwide total aquaculture production.

Europe and the region formerly comprising the USSR together account for another 10 percent, about one-third from the former USSR, another one-third from Spain and France, and the rest scattered throughout the region. The African continent produces only 0.5 percent of the total and the entire Western Hemisphere less than 5 percent. The U.S. contribution to world aquaculture of approximately 0.3 mmt equals only about 2 percent of the total.

Algae (seaweeds)—grown for both food and chemicals (agar, alginic acid, and carrageenan, used as stabilizers and emulsifiers in the food, cosmetic, and pharmaceutical industries)—are the leading marine aquaculture product by weight, yielding some 4 mmt per year. Mollusk farming produces 3 mmt, about equally divided among oysters, clams, and mussels, with smaller quantities of scallops. The culture of marine crustaceans is restricted to shrimp or prawns *(Penaeus* spp.), a rapidly growing industry worldwide.

Of the 7 mmt of finfish produced in 1988, 6 mmt represented freshwater species, including carp and tilapia grown mostly in Asia. Less than 1 mmt of marine finfish were produced, including roughly 200,000 metric tons each of milkfish, Japanese yellowtail (amberjack), and salmon.

The monetary value of the 1988 world aquaculture crop of 14 mmt was estimated at $22.5 billion, an increase of 19 percent from the $18.8 billion value of the 1987 crop and nearly twice that of the 1985 yield ($13.1 billion) (FAO, 1990). The values are undoubtedly underestimates because only 60 of the 144 countries that now report statistics to FAO include information on prices and values.

STATUS OF U.S. MARINE AQUACULTURE

Of the roughly 0.3 mmt of aquatic life grown in the United States, nearly three-quarters are freshwater organisms. Most of the freshwater production consists of catfish, crayfish, and rainbow trout, in that order of importance. Large numbers of freshwater organisms are grown for purposes other than their immediate use for food. These include ornamental fish, baitfish, trout, and other species stocked for recreational fishing.

Marine aquaculture is dominated by oyster culture (80 percent of the total), which is, however, a declining industry in the United States. Clams, mussels, salmon, and shrimp make up the remaining 20 percent, in order of importance. The technology is currently being developed for a few other marine species (e.g., abalone, red drum, scallops, striped bass, and white sturgeon), but as yet they are produced commercially in insignificant quantities. The production and monetary value of the various U.S. aquaculture crops are summarized in Table 2-2. In both categories the United States is equal to about 2 percent of world totals.

Domestic consumption of fish products grew in the 1980s primarily be-

TABLE 2-2 U.S. Aquaculture Production, 1988

	Production (metric tons)[a]	Value ($ million)[a]
Freshwater		
Catfish	155,000	265
Crayfish	30,000	25
Trout	25,000	65
Striped bass (hybrids)	450	2
Bait/ornamental fish	—	75
Alligators	—	20
Subtotal	210,450	452
Marine		
Oysters	63,000	50
Clams	8,000	10
Mussels	4,000	2
Salmon	3,000	22
Shrimp	1,000	3
Subtotal	79,000	87
Total U.S.	289,450	539
Total world	14,000,000	22,500
U.S. as percentage of world total	2.0	2.4

[a]Figures are rounded.

SOURCE: Compiled from U.S. Department of Commerce (1990) and Food and Agriculture Organization (1990).

cause of the recognition of the health attributes of fish relative to other meat products, the strong U.S. economy, and rising real per capita incomes. Real per capita disposable income rose 16.6 percent between 1980 and 1988, and real total personal disposable income rose 26.1 percent (Council of Economic Advisors, 1989). Per capita consumption of fish products in the United States rose 24 percent from 12.5 pounds per capita (retail weight) in 1980 to 15.5 pounds per capita in 1990 (see Figure 1-1).[2] The last few years have shown more or less stable per capita consumption despite the fact that prices for fish are increasing faster than for meat and poultry products. From 1980 to 1990 the consumer price index (CPI) for fish increased by almost 68 percent, from 87.5 to 146.7 (CPI base year 1982–1984). This figure compares to increases in the CPI of 38.6 percent for meat, 41.4 percent for chicken, and 52.5 percent for all foods (Putnam and Allshouse, 1991). The increase in per capita consumption, combined with the sharp rise in the relative price for fish, has resulted in steadily increasing expenditures for seafood (see Figure 1-2) and indicates a shift in consumer preferences toward seafood.

As a major seafood-consuming nation, the United States has remained dependent on imports for between 64.7 (1986) and 43.3 percent (1990) of edible supplies over the past decade. The recent improvement in domestic supply share reflects the large increase in Alaska's landings of the expand-

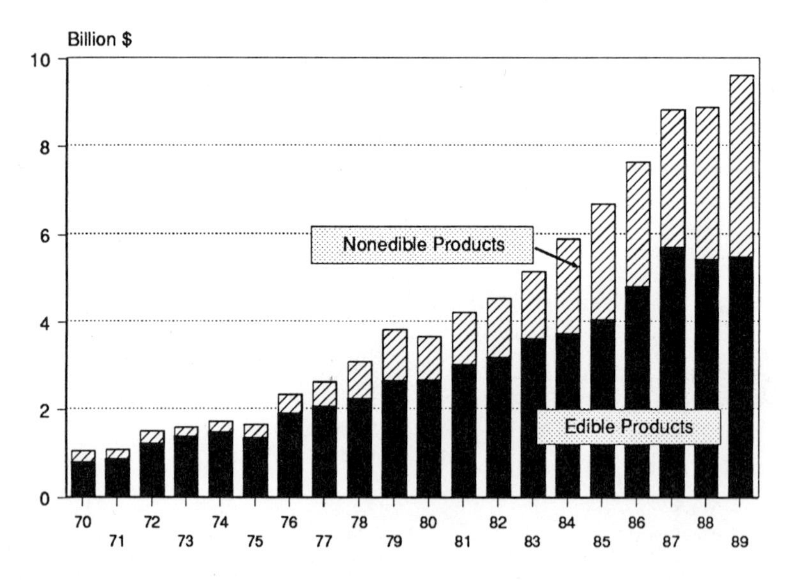

FIGURE 2-1 U.S. trade in fishery products: value of imports, 1970–1989. SOURCE: Compiled from U.S. Department of Commerce, *Fisheries of the United States, 1970–1990* (various issues).

ing pollock fishery. The 1989 import level remains impressively high at $9.6 billion (see Figure 2-1).

The trade deficit in edible fishery products alone has risen from approximately $1.8 billion in 1980 to $3.2 billion in 1989. The trade deficit increased from $2.6 billion in 1980 to $5.5 billion in 1990 (see Figure 2-2) if nonfood fishery products are included (e.g., jewelry, live trout, live eels, ornamental fish, feed, vitamins, agar, seaweed, reptile skins, fur-derived products, and other products).

It is useful to compare the magnitude of fishery imports with traditional agricultural products. As can be seen from Figure 2-3, in 1989 imports of fishery products exceeded those of all traditional animal products as well as the sum of all horticultural products, all grains, and all "noncompetitive" products, which include coffee and bananas. Shrimp imports alone are in the range of the value of all beef imports, all wine and beer imports, and all fruit and vegetable imports.

The contribution of marine aquaculture to imports continues to increase. Both cultured shellfish (primarily shrimp) and cultured finfish (primarily salmon) are imported from approximately a dozen geographically diverse

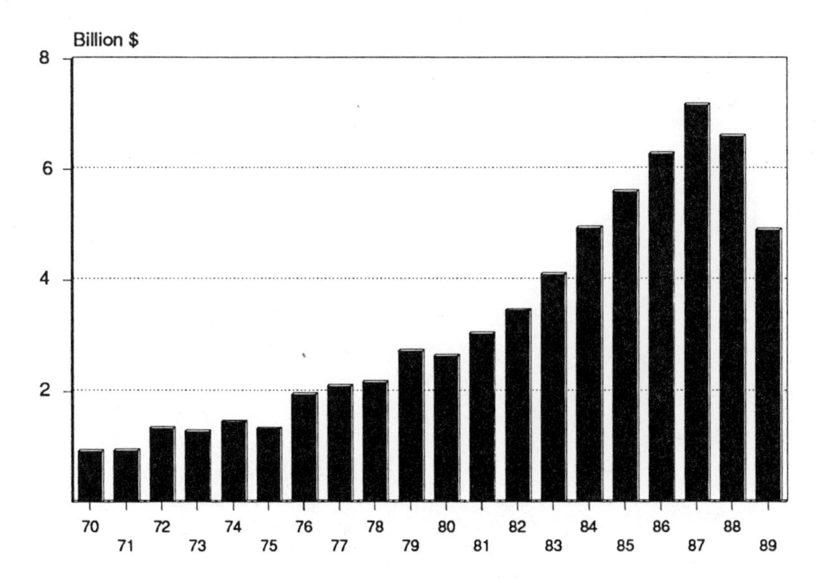

FIGURE 2-2 U.S. trade deficit in fishery products, 1970–1989. NOTE: In 1989, the definition of "nonedible" fishery products was broadened to include many additional manufactured products previously not included, which are exported by the U.S. This change explains much of the decline in the deficit for that year. SOURCE: Compiled from U.S. Department of Commerce, *Fisheries of the United States, 1970–1990* (various issues).

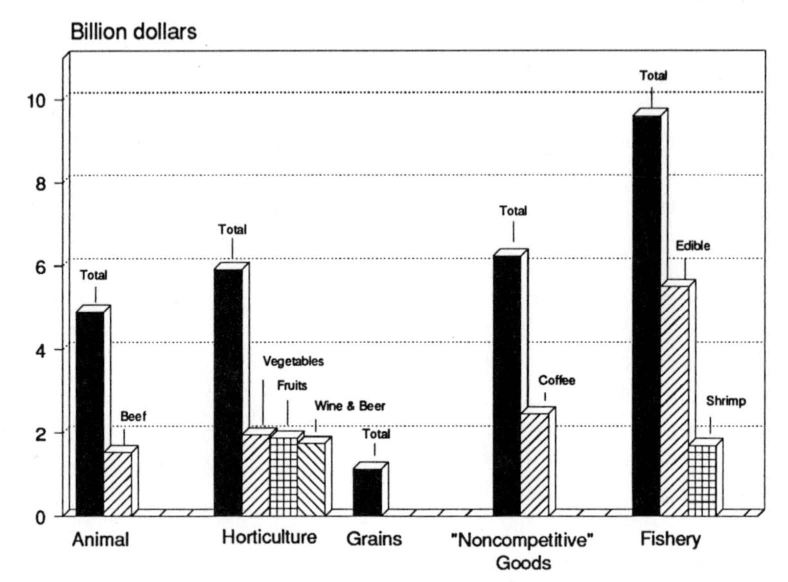

FIGURE 2-3 U.S. agricultural and fishery imports, by categories, 1989. NOTE: Agricultural commodity fiscal year, as well as the fishery product calendar year. SOURCES: U.S. Department of Agriculture (1990); *Outlook for U.S. Agricultural Exports*. U.S. Department of Commerce (1990); *Fisheries of the United States*.

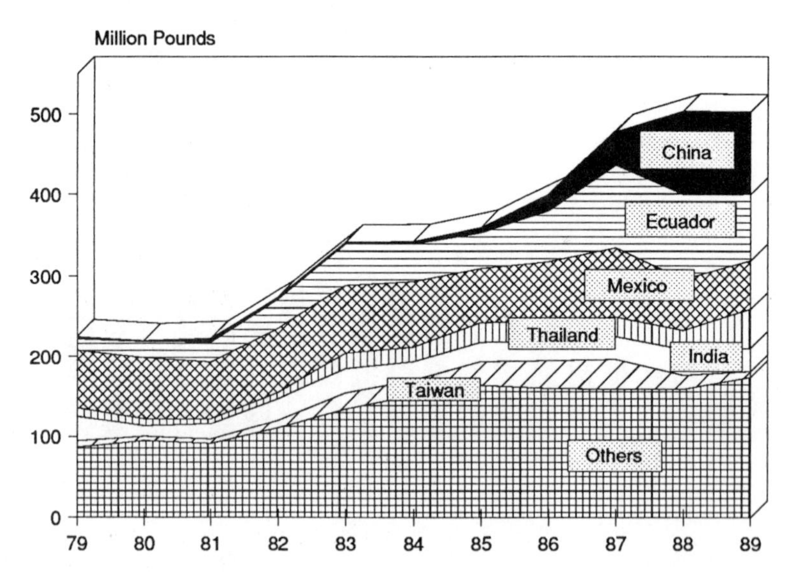

FIGURE 2-4 U.S. imports of shrimp, by country of origin, 1979–1989. NOTE: Production in China, Ecuador, Thailand, and Taiwan is dominated by aquaculture. SOURCE: Compiled from U.S. Department of Commerce, *Fisheries of the United States 1980–1990* (various issues).

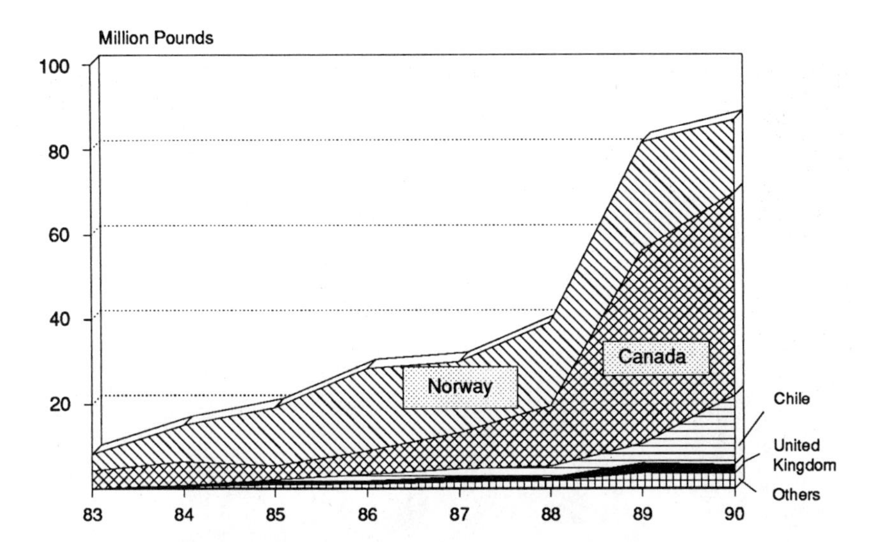

FIGURE 2-5 U.S. imports of fresh salmon, by country of origin, 1983–1989. NOTE: With the exception of Canada, virtually all of these imports are from aquaculture. SOURCE: U.S. Department of Commerce, *Import Statistics* (various issues).

countries (Figures 2-4 and 2-5). Substantial quantities of seafood come from Scandinavia, South America, Central America, and Asia. Salmon aquaculture imports have increased steadily despite the fact that the United States is the largest producer of salmon from capture fisheries in the world. However, a recent ruling by the U.S. International Trade Commission against Norway (see discussion in Chapter 3) has dramatically reduced imports of Norwegian salmon.

Marine aquaculture of finfish in the United States is currently an embryonic and struggling industry. Most of the success to date has been with salmonids: in particular, coho, chinook, sea-run rainbow trout, and Atlantic salmon on the West Coast, and Atlantic salmon and sea-run rainbow trout on the East Coast. A number of fledgling and experimental operations are attempting to culture other species: hybrid striped bass in the mid-Atlantic, Southeast, and Southwest; red drum in the Southeast; dolphin (mahi mahi) and ornamental marine tropical fish in Hawaii, and freshwater culture of anadromous sturgeon and striped bass in California.

Except for salmonid culture, the marine finfish aquaculture industry is relatively small. Consequently, for most other species, few data are available on the number of firms, employment, revenues, and quantity produced. Data are collected only sparsely by government agencies, and many firms'

Red drum *(Sciaenops ocellatus)* harvested from an experimental intensive culture pond in South Carolina.

business lives are short. Following is a review of the major marine species presently under culture in the United States.

Mollusks

Oysters

Culture of the American oyster (*Crassostrea virginica*) is the oldest form of marine aquaculture practiced in the United States. The species occurs along the entire eastern U.S. seaboard from Maine to Florida and throughout the Gulf of Mexico. Virtually all oyster production in these geographical areas involves some human intervention and manipulation, however primitive, and is therefore a form of aquaculture.

The industry has been in steady decline for more than 70 years, from a peak production of 0.25 mmt in 1920 to about one-tenth that amount today. Chief among its problems are overfishing and habitat loss, as well as a series of uncontrollable disease epidemics, one of which has almost eliminated oysters from the northern part of their range. Pollution has had a devastating impact on oyster cultures in the San Francisco and Chesapeake bays. Another serious constraint is the closure of shellfish beds for public health reasons because of human pollution and/or blooms of toxic unicellular algae (red tides) (Virginia Sea Grant, 1990). Statistics indicating decreasing per capita consumption of oysters actually reflect domestic availability as well as consumer preference. Any effect of consumers' reactions to health concerns on per capita consumption was masked by a 40 percent decrease in domestic supply (USDC, 1989). In 1990, the wholesale value of the domestic oyster supply was approximately $25 million. A good opportunity exists to revitalize oyster production through new technology.

The technology for growing the American oyster is well established, although the most efficient methods (i.e., raft and rack culture—see Appendix A) are generally not allowed in most U.S. coastal waters for aesthetic or environmental reasons. Currently, most of the culture practices are limited to planting of shells or other clutch material to "catch" oyster spat, harvesting in a controlled manner to maintain desirable standing crops on beds, and transplanting seed oysters from beds in one area (often moderately polluted) to clean beds elsewhere. Little technological innovation has occurred in the last several decades.

Chronic diseases are now widespread throughout the geographical range of American oysters, threatening the continued existence of the industry. Although some progress has been made in developing disease-resistant oyster strains, much more research is needed on the prevention or curing of such diseases.

The Pacific oyster (*Crassotrea gigas*) introduced from Japan is the primary species cultivated along the Pacific Coast. Some populations have established themselves and spawn naturally, but little use is made of their seed. The reason for this is the development of the remote setting process whereby oystermen have built seed-catching tanks on their own farms or have "eyed" oyster larvae shipped in from private hatcheries for setting. Although the concept of shipping eyed larvae was tried in the 1960s, it did not become a reality on a commercial scale until the late 1970s. One hatchery can produce billions of eyed larvae in any given year and they can be shipped with ease. With practice, growers have a good success rate for seed settlement on material placed in the tanks. Thus, seed production for the Pacific Coast of the United States is no longer a problem. A similar procedure began in 1990 for the American oyster when a hatchery opened in Louisiana.

Aside from health considerations arising from human and industrial pollution, there is no indication that disease is widespread in Pacific oysters cultivated on the West Coast. However, oysters in Coos Bay, Oregon and elsewhere show malformations due to toxic effects of TBT (tributylin) from anti-fouling paints (Wolniakowski et al., 1987). Furthermore, the Pacific Coast is increasing production to satisfy the market demand generated by problems of disease in oysters cultivated elsewhere. Production of the Pacific oyster in Washington was reported to be 29,378 metric tons in 1988 (Chew and Toba, 1991), exceeding that of American oyster production from the East Coast (including the once most productive Chesapeake Bay area). Gulf of Mexico production is still higher than Washington production. Limitation of submarine leases in the Chesapeake Bay is also a factor. Attempts to grow oysters in closed systems have been extremely expensive.

Clams

Clam farming in the United States is in its infancy, with most of the aquaculture production coming from the hard clam (*Mercenaria mercenaria*) and the Manila clam *(Tapes japonica)* (Chew and Toba, 1991).

The hard clam (*Mercenaria mercenaria*) is a popular bivalve that ranges along the eastern U.S. coast, with subspecies occurring throughout the Gulf of Mexico. Wild stocks of hard clams are becoming scarce, while the species has become an increasingly popular alternative to the disappearing oyster. Most valuable is the smallest legal size (2 inches long in most states) served raw on the half-shell, bringing as much as $0.25 each to fishermen or growers.

Clam farming is a new but growing industry along the entire Atlantic Coast, held back primarily by disease-related problems and regulatory con-

Newly set hard clams are grown in "upwellers" in an indoor nursery to the size of about 3mm before moving to outdoor culture systems.

Bay scallops *(Argopectin irradiens)* grown in Chinese "lantern" nets suspended from buoyed lines in a start-up commercial venture in Massachusetts.

straints on leasing and harvesting. The technology for clam culture is well developed, including hatchery production of seed; however, production of algal food for hatchery and nursery operations is a severe economic and technical constraint in most areas. Research on inexpensive replacements for algae has been largely unsuccessful.

Bottom (buried) culture, which is usually done in enclosures or under protective netting, is the only existing technique for growing hard clams.

Off-bottom culture has not proved successful, but it also has not been adequately evaluated. The demonstration that hard clams can be grown to marketable (little neck) sizes at densities of 100 per square foot of bottom in small experimental systems has led to unrealistic commercial projections (e.g., millions of clams per acre) in some cases. Environmental modeling based on available food, water circulation, and other variables is needed to be able to predict the carrying capacity of a given environment for cultured clams as well as for other bivalves.

The Manila clam (*Tapes japonica*) was introduced inadvertently with Pacific oyster seed shipments from Japan and now has grown to be a major component of the shellfish production for the state of Washington. Natural brood stocks have been established for the Manila clam, and with the increase in demand, new hatcheries have been built to produce Manila clam seed for planting on open natural beds or with clam netting over natural beds. New techniques are needed for growing this important clam species, such as using cages or shell bags.

The geoduck clam *(Panopea generosa* spp.) is also cultured for planting. These clams are spawned in the hatchery and cultivated through a nursery system, then planted back in the subtidal geoduck grounds. Further research is needed to increase the survival rate of these clams after two to three years. The wholesale value of all domestic clams was approximately $280 million in 1988 (Chew and Toba, 1991).

Scallops

Scallop culture is well developed elsewhere in the world, but it is in an early exploratory phase in the United States. The technology is much the same as for other bivalves. Seed are collected from the wild or grown in hatcheries by using the same basic methods as for clams and oysters. Seed are grown to marketable size on the bottom, in suspended lantern nets, or in other off-bottom devices. A shorter grow-out time is an advantage of most scallops over clams and oysters; usually, scallops require about one year or less from the egg. As yet, no established commercial scallop culture projects exist in the United States, although several small companies are in the start-up phase.

Mussels

The blue mussel (*Mytilus edulis*) is an extremely popular shellfish in Europe and Asia, but its use as food is just now becoming accepted in the United States, where a budding industry is developing on both coasts. Where mussels are naturally abundant, collection of wild seed (on ropes or other substrata) is so easy as to preclude the need for hatcheries. Seed are grown out on ropes where legally permissible or on cleared bottom. The major constraints to mussel culture have been a limited market and low value for the product, exacerbated by the ready availability of wild stocks. Findings of paralytic shellfish poisoning (PSP) in mussels have also dampened consumer demand. The market is expanding gradually, but its growth probably could be accelerated. Approximately 4,000 metric tons of mussels were grown in the United States in 1988 (FAO, 1990), mostly in Maine (Wilson and Fleming, 1989).

Crustaceans

Shrimp

A great deal of enthusiasm for shrimp aquaculture has resulted from commercial successes in Ecuador, Taiwan, China, Japan, and Indonesia. Particularly noteworthy has been the ability of these industries to develop significant export earnings (or reduced imports, as in the case of Japan). The United States constitutes the world's largest market for shrimp and is one of the leading countries in the development of shrimp farming technology. However, the United States lacks some of the factors contributing to the success of shrimp culture in other countries—large expanses of inexpensive and undeveloped land adjacent to estuaries, cheap labor, abundant natural supplies of postlarvae and brood stock of preferred species, year-round growing conditions, and lack of environmental regulation.

Domestic shrimp farms are relatively few (probably no more than 25 to 30 nationwide) and range in size from perhaps 1 to more than 400 acres (Chamberlain, 1991; Hopkins, 1991; Pruder, 1991). Farms are located principally in Hawaii, South Carolina, and Texas. Extensive to intensive technology[3] is employed in South Carolina. In Hawaii and Texas farms tend to be principally semi-intensive in operations, although research on intensive technologies is going on in all three states (Chamberlain, 1991; Pruder, 1991; Sandifer et al., 1991a,b). Total U.S. production of farmed shrimp in 1990 was estimated at 900 metric tons (on a head-on basis) (Rosenberry, 1991).

A number of shrimp operations have failed in the United States, but many of these failures occurred before production technology had developed to the point it is at today. In particular, technology for intensive production of marine shrimp in ponds appears to be making U.S. production more competitive with foreign shrimp farmers. Technology aimed at diminishing the disadvantages of high-cost land and labor, as well as temperate climatic factors, is necessary (Sandifer et al., 1991a,b). Major technological constraints facing domestic producers include a supply of specific pathogen-free stocks of the preferred species (*Penaeus vannamei*), reduction of biochemical oxygen demand (BOD) and nutrient loading in wastewater ef-

Shrimp jumping as an intensive culture pond in South Carolina is drained.

fluents, and the need for genetically improved stocks, better feed, and disease control.

Major factors restricting success of shrimp farming in the United States at present are

- a variety of regulatory problems, especially related to effluent waters;
- lack of a native species with preferred characteristics for culture (nearly all U.S. culture operations are based on nonindigenous species);
- limited availability of postlarvae at times of greatest demand;
- disease concerns;
- insufficiently refined technologies for maintaining and routinely reproducing completely closed brood stock populations;
- high cost of major inputs (i.e., land, labor, equipment, electricity, feed, money); and
- "softening" of prices for sizes of shrimp most readily produced by aquaculture (see Figure 2-6).

Shellfish Opportunities

A focus of aquaculture research efforts in the 1970s was the American lobster *(Homarus americanus)*. This interest was fueled by supply limits,

Raceway-recirculating system for rearing and harvesting of shrimp.

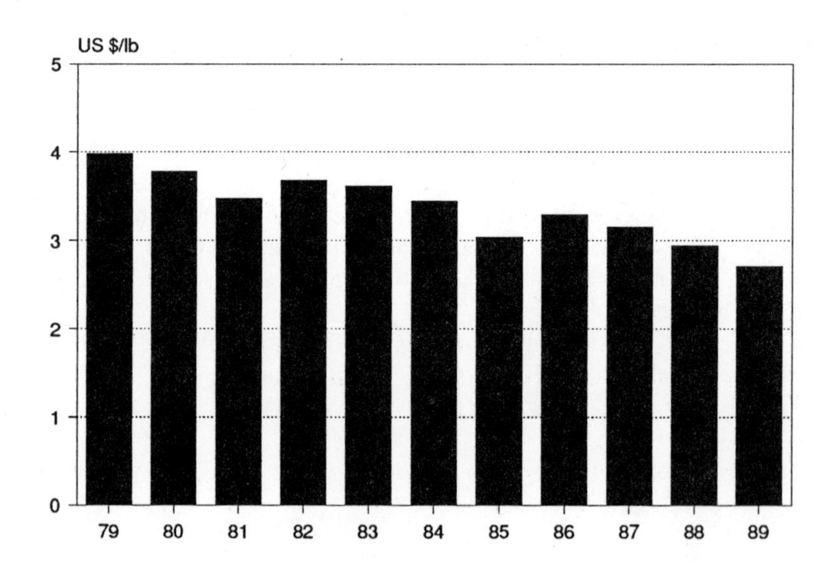

FIGURE 2-6 Real prices of U.S. imports of shrimp, 1979–1989. NOTE: Prices are adjusted by the consumer's price index for food (1982–1984 = 100). SOURCES: U.S. Department of Commerce, *Fisheries of the United States* (various issues), and the Consumer Price Index, February, 1990.

market image, and a relatively high price for the species. The American lobster defies commercial cultivation because of its aggressive nature that makes it necessary to grow each lobster separately. Although this has proved uneconomical to date, the escalating value of lobsters, particularly in foreign markets, may change the economics in the near future. Except for solving the intractable problem of cannibalistic behavior, the technology for rearing lobsters, including hatchery production of juveniles, is well established. However, a pelletized artificial feed of the proper consistency and nutritional value has yet to be developed.

Lobsters maintained at 22–24°C can be reared from the post-larval to a one-pound marketable size in less than two years (Hughes et al., 1972). From 1980 to 1989, domestic production increased almost continuously to a record 52.9 million pounds (USDC, 1990). During the same period, imports of fresh and frozen lobster tripled to 69.1 million pounds in 1989. Lobster prices have increased steadily in spite of these record landings and imports. Researchers have encountered difficulty in reducing costs. With

A nursery for rearing juvenile American lobsters *(Homarus americanus)*, which must be kept separated because of cannibalism.

producer prices of lobster ranging from $1.75 to $4.00 per pound, a technology breakthrough may be essential for the commercial culturing of this species to become economically feasible. Technologies to improve genetic selection for rapid growth and survival, and to address the cannibalistic nature of the lobster, would greatly aid economic feasibility.

The spiny lobster *(Panulirus)* of Florida and the Caribbean cannot yet be grown through its many larval stages routinely, although Japanese scientists have carried a few individuals through. Postlarval spiny lobsters (pueruli) can be collected readily as they metamorphose from the planktonic to benthic habit, sometimes in very large numbers, and these may be grown out to marketable size in 12 to 18 months in captivity with no great difficulty. The spiny lobster does not share the cannibalistic nature of the homarids and would, therefore, be preferable for commercial culture. However, collection of pueruli from the wild is unacceptable in most places because of concerns about potential impacts on natural populations and fisheries. Thus, spiny lobster culture most probably must await perfection of the technology for rearing the animal throughout its entire life cycle.

Abalone also has potential for development. The current market for abalone is not large but the product commands a good price. Companies in California and Hawaii are in the early stages of commercialization. Culture systems for abalone, unlike other mollusks, are based onshore; seawater is pumped to some form of tank or raceway facility that has an abundance of support surfaces for the animals to attach to during grow-out.

There is no true culturing of crabs, although some portunid crabs are reared in farm operations in Southeast Asia. However, large numbers of premolt blue crabs *(Callinectes sapidus)* are captured and held in shallow floating cages or tanks (both flow-through and recirculating) until they molt to the soft-shelled stage (ecdysis), at which time they are harvested and sold as a delicacy. The technology is well established and simple, and the industry is growing; however, improvements in holding and "shedding" systems, especially recirculating ones are needed. The greatest restriction is the supply of premolt crabs from the wild. Research is focusing on ways to easily and inexpensively stimulate ecdysis in large groups of crabs more or less simultaneously.

Finfish

Salmon

Floating cage or net-pen culture (see Chapter 5) of growing Pacific salmon to "pan size" over one season (i.e., one season postmolt) originated in the Puget Sound area of Washington in the early 1970s (Naef, 1971). Never

Raceway system for onshore production of abalone (the Abalone Farm, Cayucos, California). The raceways contain a maze of support surfaces onto which the abalone attach. Seawater is pumped to the facility; seaweed is harvested from the ocean and placed in the raceways to feed the animals. The roughened water surface is caused by the water aeration system.

highly successful, that practice has now all but disappeared. During the 1980s however, the Norwegians began cage culture of Atlantic salmon over two growing seasons (i.e., 18 months from smolt) to produce fish that averaged 4 to 5 kilograms (kg). This practice proved highly successful and spread quickly around the world, including the United States and Canada. Foreign ventures have often been started by Norwegians, who were constrained by law from further expansion within their own country.

The United States has limited availability of suitable coastal farm sites. However, undeveloped sites exist in Maine; Washington has numerous sites that would be suitable for salmon farming in the Puget Sound area; and

Alaska has literally thousands of miles of coastline that would be ideally suited for the development of salmon farming. In Washington, Alaska, and Maine, the development of salmon farms is impeded by local and state regulations. The Alaska legislature has mandated a moratorium on the development of private, for profit salmon farming. Washington has developed a complicated but orderly process for licensing and regulating salmon farms in the Puget Sound area, but the costs of compliance are substantial, and they have tended to discourage investment in the development of salmon farms in the area. For the above reasons, many entrepreneurs, including Norwegian and other foreign investors, who initially attempted to establish salmon farms in Maine and Washington, moved north to British Columbia

Atlantic salmon and rainbow trout cage culture in Maine.

Atlantic salmon net cage culture in Norway.

and New Brunswick where legal constraints were less onerous. By the late 1980s, there were 175 operational salmon pen culture sites in Canada and 26 in the United States (Bettencourt and Anderson, 1990).

Although most salmon farms involve culture in cages, some limited production occurs in tanks or raceways located onshore. One such operation in California pumps seawater into tanks located a short distance inland.

In 1989, Norway, the acknowledged leader in salmon aquaculture, produced 116,164 metric tons of farmed Atlantic salmon (FAO, 1991). Because this followed an unusually productive season for farmed salmon (see Figure 2-7), as well as for wild-caught salmon, the price dropped precipitously (see Figure 2-8) to a level that was, in most cases, below the cost

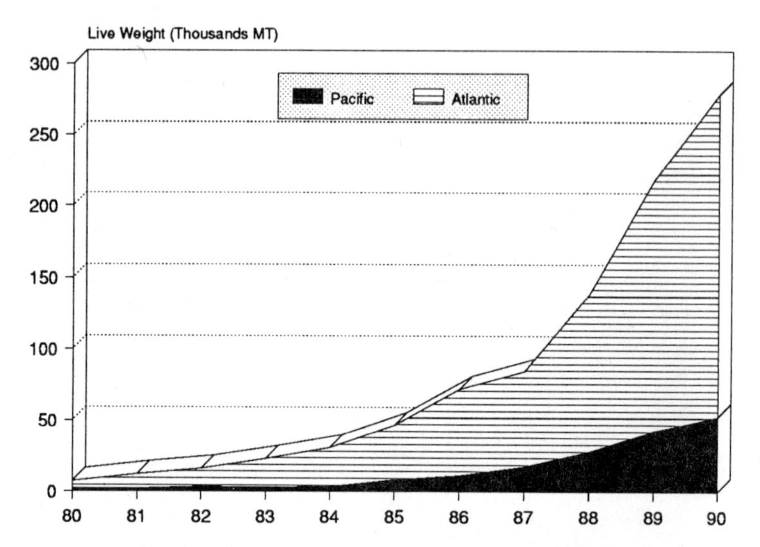

FIGURE 2-7 World supply of farmed Atlantic and Pacific salmon, 1980–1990. SOURCES: Compiled from 1980–84: U.S. Department of Commerce *Import Statistics* (various years). 1985–1989: Food and Agriculture Organization. 1989–1990: estimated.

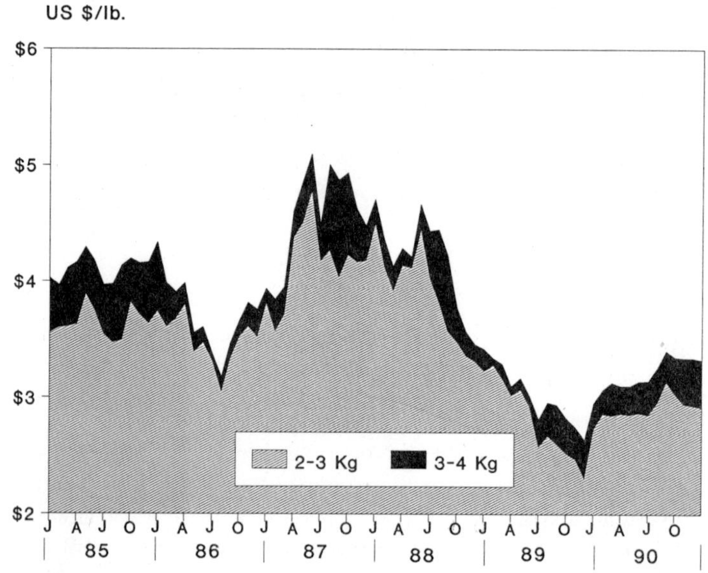

FIGURE 2-8 Price variation of fresh Norwegian mid-Atlantic salmon, 1985–1990. NOTE: Prices are adjusted by the monthly consumer price index (1982–1984=100). Prices are for first receivers in the mid-Atlantic region, for whole, head-on fish. SOURCE: *Urner Barry Seafood Price: Current* (various issues); *Economic Report of the President* (various issues).

of production and that drove several companies out of business almost immediately. The industry had not yet recovered at the time this report was written.

Before the salmon market crash of 1989, most U.S. farms were just beginning to approach profitability. Current performance of the industry is therefore difficult to assess. In 1989, farmed salmon and steelhead pro-

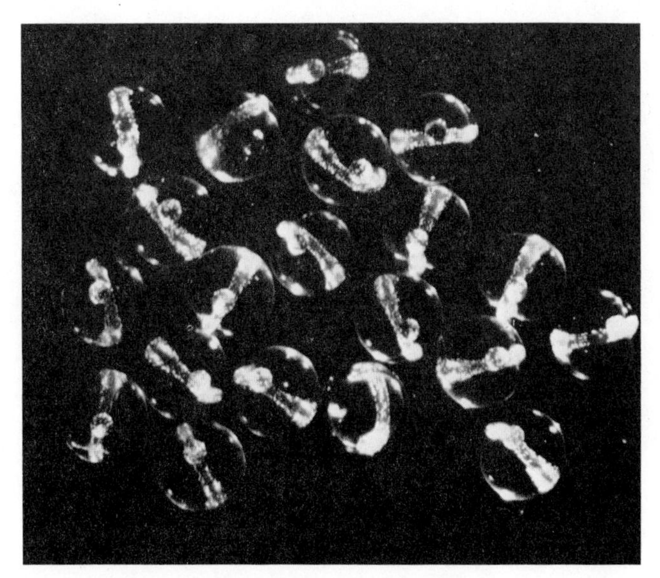

Red drum eggs, approximately 15 hours old.

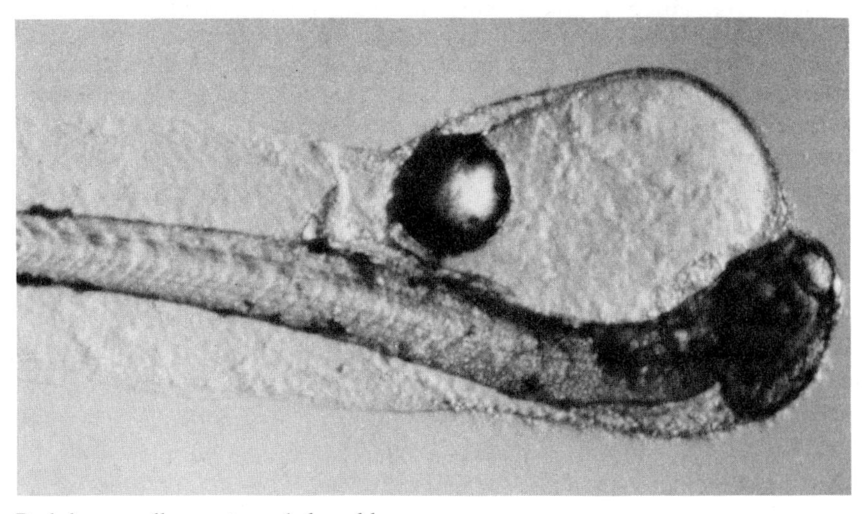

Red drum, yolk sac stage, 1 day old.

Red drum adult in spawning tank.

duction in the northwestern United States was reportedly 2,309 metric tons (J. Pitts, Market Development Division, State of Washington Department of Agriculture, personal communication, 1991), and Maine produced an estimated 1,440 to 1,650 metric tons (Bettencourt and Anderson, 1990).

Red Drum

Red drum (*Sciaenops ocellatus*), which is known also as redfish, is common in south Atlantic and Gulf of Mexico waters, where it is esteemed highly by sports fishermen and consumers. A great deal of interest in the culture of red drum has developed in recent years owing to increased consumer appeal, implementation of fishing restrictions (particularly limiting commercial take), and development of technology to produce fingerlings for stock enhancement programs. Techniques for captive reproduction and fingerling production are fairly well established (Chamberlain et al., 1990). Approaches to grow-out have been developed both for pond and for tank culture.

At present, interest in establishing commercial red drum farms is focused primarily in the south Atlantic and Gulf states. Experimental grow-out in South Carolina yielded harvests of 9,000 to 24,000 kg per hectare of 1-kg fish, depending on stocking density (Sandifer et al., 1988). The grow-out period was approximately 18 to 20 months for a 1- to 5-kg fish, but the production of 1-kg fish in approximately 12 months in a more tropical environment is projected (Sandifer, 1991).

A commercial farm on Galveston Bay, in Baycliff, Texas, specializes in the closed-system tank culture of red drum. Temperature and photoperiod controls produce eggs on demand. Larvae are stocked at densities of 200,000 per 300-gallon tank. After being weaned to dry food, fry are grown to fingerling size in 1,500-gallon tanks and 6- to 8-inch fingerlings are then reared in 27,000-gallon tanks for final grow-out to 1-kg marketable fish. Fingerlings stocked in spring are ready for harvest by October or November, thereby avoiding lethal winter temperatures. Grow-out tanks are closed, recirculating systems using mechanical particulate filters, biological (biodisc) filters, and oxygen injection (Holt, 1992).

During peak commercial production from natural fisheries, the wholesale red drum market was approximately $30 million. Development of culture techniques can serve to bring this food fish back to consumers.

Production facility (the Fishery, Sacramento, California) for culturing striped bass, white sturgeon, and catfish. Hatchery operations are sheltered by the open shed (right center). Sturgeon and some other fish are reared in tanks (foreground), and striped bass and catfish are reared in ponds (barely visible beyond the buildings). Water discharge from the tanks flows to the ponds.

Red snapper *(Lutjanus campechanus)* in spawning tank.

Other Marine Finfish

Of the various warmwater marine finfish species, striped bass (*Morone saxatilis*), white sturgeon *(Acipenser transmontanus)*, and hybrids with its freshwater cogeners (*M. chrysops, M. mississippiensis*) appear to have the greatest near-term potential for commercial development in the United States. In general, the hybrids have proved to be more hardy and otherwise more suitable for aquaculture. The striped bass itself is anadromous, and both it and its hybrids grow equally well, if not better, in hard freshwater than in seawater. All of the early commercial ventures at growing hybrid striped bass are based on freshwater systems. However, some people believe that cultivation in coastal salt or brackish water ponds will ultimately prove more successful, both technically and economically, than the initial commercial efforts to grow the hybrids intensively in tanks using pumped geothermal freshwater (Doroshov, 1985). Both striped bass and white sturgeon have been shown to have potential for commercialization in California. Both are anadromous and have been cultured successfully in tanks and ponds using fresh water.

Although typically reared in fresh water, saltwater culture of some species of tilapia and their red hybrids is now showing economic promise in several countries, including the United States. In Hawaii and the Caribbean, the growth of a red hybrid was found to be significantly greater in cages placed in brackish water shrimp ponds than in freshwater ponds (Meriwether et al., 1983, 1984).

The dolphin mahimahi (*Coryphaena hippurus*) has been spawned and reared in captivity in Hawaii and has exhibited impressive growth rates, reaching 1.3 kg in 130 days from hatch (Hagood et al., 1981). Regular spawning in captivity has been demonstrated (Kraul, 1992) and is essentially routine. The species is pelagic and piscivorous, however, which suggests that a high amount of natural marine foods would have to be incorporated into its diet. Nutritional studies confirmed the high requirement for animal foods but indicated that a substantial portion (perhaps as much as 50 percent) could be replaced with much less expensive plant-based foods such as catfish feed (Szyper et al., 1984).

A number of other marine finfish species that are harvested in capture fisheries in U.S. waters are attractive candidates for marine aquaculture. These include halibut, swordfish, shark, flounder, sole, cod, rockfish, pompano, snapper, grouper, and weakfish. Production of these species, however, will require development of new technologies. Many species require a long growing period to produce a marketable product. In addition, economical hatchery and grow-out techniques have not been developed for most of the species, many of which have complex early life histories involving one or more metamorphoses between life stages. Although some of these species have been cultured successfully in the laboratory, additional research is required to develop economical methods for their artificial propagation (Tilseth, 1990).

Algae

Macroscopic Algae (Seaweeds)

Seaweeds are grown commercially in China, Japan, Taiwan, Korea, and the Philippines, both for human food and for extraction of the polysaccharides agar, algenic acid, and carrageenan. Depletion of wild stocks, particularly of agarophytes, has enhanced the value of these seaweeds to $1,000 or more per dry ton and has made their cultivation more attractive. However, no commercial seaweed farms for polysaccharides currently exist in the United States. A small commercial project is under way near Halifax, Nova Scotia, for cultivation of *Chondrus crispus* (Irish moss), a carrageenan source.

The most popular and valuable edible seaweed is *Porphyra* (nori), grown extensively in Japan. A state-supported research project in Washington, in which the Japanese technology was closely followed, led to initial start-up

Algae *(Tetraselmis chuil)* produced as food for rotifers, which are in turn fed to larval fish.

of several small commercial nori culture projects in the Puget Sound area. Objections to these raftlike operations, on aesthetic or environmental grounds, have caused most if not all of them to close down or move (mostly to British Columbia). Nevertheless, the current importation of more than $50 million worth of nori from Japan for Asian populations in the western United States and Canada, and for increasingly popular "sushi bars" throughout the country, makes culture of this seaweed commercially interesting.

Unicellular Algae (Phytoplankton)

Several species of unicellular algae are grown routinely throughout the world as food organisms for larval and juvenile mollusks and crustaceans. Culture systems range from tanks and cylinders in hatcheries to outdoor ponds an acre or more in size. Although small-scale culture (100 gallons or less) has become routine, large-scale algal culture, particularly in outdoor ponds, has proved difficult for two major reasons: (1) inability to control the species composition in culture and thereby to prevent undesirable species from taking over, and (2) predation from microcrustaceans, protozoans, and other animals accidentally but inevitably introduced into the system.

Algal culture is probably the most difficult and costly component of shellfish hatchery operations. The high cost of harvesting microscopic algae from water by centrifugation, filtration, or flocculation has always made algae culture economically problematical for low-value products (e.g., feed), but production of high-value chemicals has changed the economic picture. For example, certain unicellular algae contain pigments or other fine chemicals of high value. A commercial firm in Hawaii and two in southern California currently grow two species (the blue-green *Spirulina* spp. and the flagellate *Dunaliella salina*) for such products, most of which are exported to Japan. These algae have unusual environmental requirements—high carbonates for *Spirulina* and hypersalinity for *Dunaliella*, both of which deter contamination from other algae and predators. The chemicals required are costly to replenish, so a recirculating system is used for their cultivation.

A few other small commercial marine aquaculture projects produce unicellular algae in the United States, but their status is not known.

MARINE FISHERIES ENHANCEMENT

Marine fisheries/stock enhancement is the release or stocking of hatchery-reared juvenile fish, mollusks, crustaceans, or other organisms into a natural marine environment where they will supplement the existing population and thereby expand opportunities for harvesting, rebuilding declining populations, or establishing new populations. These activities take two forms: (1) mitigation for the purpose of replacing natural regeneration that has been destroyed by human development such as dams, and (2) enhancement for the purpose of augmenting natural runs that are overfished or declining naturally. Effective public and private efforts can contribute to replenishing of endangered and threatened species as well as commercially and recreationally important ones.

Historically, the practice of fisheries enhancement dates back to before the turn of the century, when the Bureau of Commercial Fisheries released countless thousands of newly hatched larvae of several species of commercially important marine fish in a vain attempt at augmenting natural stocks. The practice of stock enhancement was initially conceived as an attempt to mitigate the loss of natural reproduction due to overfishing, the construction of dams (which prevented anadromous species from reaching their breeding grounds), and water pollution. However, the early practice of simply releasing newly hatched fry was soon recognized as ineffective, and it was discontinued.

Despite the initially disappointing results, procedures and technology have been developed over time to the point where several species of anadromous and marine fish and invertebrates are now reared in hatcheries and the young released to the environment in attempts to enhance declining populations of commercial, recreational, and endangered species. The hatchery production of juveniles for such purposes is clearly aquaculture; however, their subsequent growth within and harvest from the natural environment following their release cannot be so designated. Once they are released, no further human manipulation or control is involved, and the fish frequently become indistinguishable from wild fish sought by capture fishermen.

The most widely recognized enhancement and mitigation activities are through federal (U.S. Fish and Wildlife Service and the National Marine Fisheries Service) and state hatcheries in producing juvenile fish for stocking in public waters (both fresh and marine) to rebuild and augment fish stocks where populations have declined due to over-exploitation, habitat loss or degradation, or a combination of the two. The agencies also are involved in introducing new species or strains into U.S. waters, including walleye and northern pike in several states outside their native range and striped bass on the West Coast.

Current planting of fish in public waters (fresh and marine) by federal and state hatchery systems exceeds 40 million pounds per year. By far the largest of these efforts in the marine environment involves Pacific and Atlantic salmon, with significant public efforts on both the Pacific and Atlantic coasts of the United States to mitigate extensive losses from development activities such as hydropower, fishing, logging, mining, agriculture, and urban growth (Nehlsen et al., 1991). Adult fish are strip-spawned, the eggs are incubated, and the larvae are reared in hatcheries. The juvenile fish are reared in fresh water to the size of smolts, the stage at which they undergo the physiological change that enables them to live in saltwater, after which they are released. Federal, state, and private nonprofit hatcheries from California to Alaska now release upward of one billion Pacific salmon smolts annually.

Due to the historically large public role in production of salmon and other species, fishery enhancement and the rehabilitation of stocks of threatened or endangered species have been traditionally considered as responsibilities of public agencies (McNeil, 1988). However, Oregon allows private entrepreneurs to produce and release smolts and then recapture a portion of the salmon that return for their own use. Currently 12 private salmon hatcheries in Oregon have permits for private ocean ranching, as the practice is called. Of the 12, 3 companies have significant operations but none has achieved profitability based on ocean ranching. In Alaska, private, nonprofit ocean ranching is practiced by hatcheries owned and operated by fishermen's cooperatives to enhance the commercial fishery (Mayo Associates, 1988).

One of the most important biological characteristics of salmonids that makes them excellent candidates for stock enhancement or ocean ranching is that the fish have a strong instinct to return to their natal stream (or point of release) upon reaching reproductive condition. Initially, the return of salmon released from hatcheries was less than 2 percent, but in northern latitudes, returns as high as 15 percent have been achieved (McNeil, 1988). Production of larger and healthier smolts through advances in husbandry techniques, nutrition, and genetic selection has contributed considerably to increased survival and return rates.

The situation with regard to stocking of salmon in public waters is very complicated, because some stocking is carried out by public agencies and other by private entities that generally expect some type of return on their investment. In the last 20 years, a number of enthusiasts have used the available salmon propagation technology developed by state and federal hatcheries to encourage investment in private ocean ranching of salmon. However, the lack of clear ownership of the fish has been one of the pri-

State of California's Nimbus Salmon and Steelhead Hatchery. Nursery ponds are used for the culture of salmon and steelhead. These fish are planted into the Sacramento River for migration to the sea.

The fish eventually return to the hatchery and swim over twenty steps to the top of the fish ladder to the holding pond where they spawn. The Pacific salmon die after spawning; all steelhead are returned to the river.

mary problems with private for-profit ocean ranching. The fact that the salmon rancher cannot recover any compensation from commercial or sport fishermen that intercept these fish before they return to their release points has contributed to the economic collapse of most operations. According to Anderson and Wilen (1986), the lack of well-defined property rights of the culturist, in conjunction with a common-property fishery, will generally result in unprofitable salmon ranching.

In contrast, the Alaskan program of private, fishermen-owned, not-for-profit salmon hatcheries has been successful. Research (Boyce, 1990) indicates that greater enhancement will probably yield additional economic benefits. One aspect of the Alaskan success is the relatively well-defined ownership of the returning salmon. The salmon are either caught by the commercial fishermen or returned to the commercial fishermen-owned cooperative hatchery.

Although ocean ranching is perceived by many as producing public and commercial benefits in restoring declining or threatened species, some researchers believe that negative interactions with hatchery fish can lead to hybridization, competition, and disease in native populations, and recommend that efforts be focused on different strategies to protect them. Among the recommended strategies are the conservation of ecosystems to allow natural reproduction of wild stocks and providing protection for certain species under the Endangered Species Act (Nehlsen et al., 1991). A new study of these issues is under way by the National Research Council.

Two other anadromous fish for which there are significant stock enhancement efforts are the striped bass and some species of sturgeon. Augmentation of freshwater and some estuarine populations of striped bass on the East Coast of the United States became routine in the 1960s, following pioneering hatchery development work at South Carolina's Moncks Corner hatchery (later the Dennis Wildlife Center) (Stevens, 1984). A similar attempt has been made to establish a Gulf of Mexico spawning stock by repeated releases of hatchery-reared juveniles into the Mississippi River system.

The species was also introduced into San Francisco Bay, California, in 1879 and 1881. Within 10 years, a major fishery developed and the population continues to support a popular sports fishery today. When numerous water and power projects began to interfere with spawning of striped bass, the California Department of Fish and Game (CDFG) initiated stock enhancement efforts that continue to the present. The species has now expanded its range from southern California to the Columbia River in Oregon.

Beginning in 1982, private producers were authorized to receive permits from CDFG to collect wild striped bass broodstock. By 1984 the demand for yearling fish to meet mitigation requirements in California exceeded the CDFG facility's capacity, so private producers were contracted to produce yearling striped bass for release into public waters. From 1982 to 1989, the number of active broodstock permittees increased from 1 to 10, and the number of adult striped bass collected from 26 to 299. In this period, the number of striped bass reared each year reached 1.5 million yearlings/fingerlings, which were sold to the State Department of Water Resources and the Pacific Gas and Electric Company for use in fulfilling part of their mitigation requirements. In addition, aquaculturists stocked 147,500 yearling bass as mitigation for the 1,475 adult bass collected for spawning. Currently, annual mitigation needs of state and private development are for 1 to 2 million yearlings/fingerlings. Overall, the striped bass enhancement program in California involving private aquaculturists has been a success, although a few anglers express concern about damage to spawning migration and disruption of fishing.

In recent years, landings of striped bass from the Chesapeake Bay and surrounding regions have decreased substantially. This decrease was believed to be due to combined effects of overfishing, habitat loss, water pollution, and disease. The difficulty of obtaining ripe broodstock was also a contributing factor. Consequently, a fishery ban was implemented in conjunction with expanded stock enhancement and research efforts. During the past 5 to 6 years, substantial numbers of tagged juvenile striped bass have been released. In 1990, the young-of-the-year juvenile index indicated that the stock had recovered substantially, and restricted levels of commercial and recreational harvest were allowed. However, the juvenile index again declined after reopening to limited fishing, so stock enhancement efforts are likely to continue.

The striped bass population in the Gulf of Mexico has never been as large as that of the Chesapeake Bay. Still, abundance of striped bass in the Gulf of Mexico has been depressed for a number of years, which has led to continuing efforts to restore these stocks via hatchery releases. However, a strong positive impact of such releases has not yet been detected.

Large-scale commercial exploitation of North American sturgeon began around 1860 and by the turn of the century most stocks had suffered drastic declines. Early efforts were undertaken to maintain the fisheries through stock enhancement, but due to a variety of problems, including the difficulty of obtaining ripe brood stock and disease, all efforts were abandoned by about 1910 (Harkness and Dymond, 1961). During the past 10 years, a number of small-scale stocking efforts have been initiated with native sturgeon and paddlefish, including the Atlantic sturgeon, *Acipenser oxyrhynchus*; the shortnose sturgeon, *A. brevirostrum*; and lake sturgeon, *A. fulvescens*; and the paddlefish, *Polyodon spathula* (Smith, 1986; Smith and Jenkins, 1991). Most efforts have been initiated recently and results to date are only preliminary. However, stock enhancement efforts with sturgeon in the former USSR appear to be highly successful (Binkowski and Doroshev, 1985). Further, populations of the white sturgeon *(Acipenser transmontanus)* in California, which support an important recreational fishery, have been augmented via a program of the California Department of Fish and Game.

The salmon, striped bass, and sturgeon discussed above are all anadromous species whose juveniles must be reared in fresh water. Hence, their culture for stock enhancement purposes is technically freshwater aquaculture, although the fish themselves may be released into brackish or marine waters. The only marine fish that is hatchery reared and released in large numbers for stock enhancement purposes at this time in the United States is the red drum, *Sciaenops ocellatus* (also known as redfish, spottail bass, and channel bass). Stock enhancement efforts with this species have been going on in Texas since 1983 (McCarty et al., 1986).

In the past several years, other states including Alabama, Florida, and South Carolina have initiated small-scale stocking efforts for red drum. The development of culture techniques and the fact that stocked red drum tend to grow rapidly and remain in the general stocking areas for the first several years make red drum good candidates for stock enhancement efforts.

A number of other marine species could also potentially benefit from aquaculture-based stock enhancement efforts. These include haddock, cod, mullet, flounder, and red snapper for commercial use and snook, tarpon, white sea bass, and spotted sea trout for recreational fisheries (Sandifer et al., 1988). Hatchery techniques for the routine mass production of these and other species need to be developed. Once hatchery methods for mass production of juveniles are established, these techniques could be used for commercial aquaculture production as well.

Stocks of molluscan shellfish also are enhanced artificially (Manzi, 1990). Virtually all oyster-producing states have some sort of enhancement program, ranging from the very simple to the complex. At the simple end of the spectrum are state requirements for planting of shell or other clutch material (i.e., material that serves as settling and attachment substrate for oyster spat, as they settle from a planktonic to a sessile, benthic existence) on bottoms each year to replace the shell removed in oyster harvesting operations and to increase the amount of suitable habitat for oyster settlement. At the other end is the production and stocking of hatchery-reared seed onto prepared bottoms in public waters (these bottoms generally are leased to private concerns).

The use of hatchery technology is widespread in the West Coast oyster industry, of moderate significance in the Northeast, and just becoming established in the Gulf and south Atlantic states (Manzi, 1990). Hatcheries are believed to be the future of the oyster industry. Stocking of hatchery-reared clams is fairly widely practiced in the Northeast and Northwest, and the largest hatchery-based clam farm in the world is developing in the Southeast. Hatchery-reared scallops also are stocked in the wild in some northeastern states.

Another form of enhancement sometimes is used to improve recreational and commercial oyster grounds. Large numbers of oysters are moved, either by hand or by machine, from marginally or moderately polluted or nearly inaccessible areas to established recreational shellfish grounds where, after an appropriate period of self-cleansing (depuration), the stocked grounds are opened for harvest by recreational gatherers. In addition, in some states, hard clams may be harvested from polluted beds, processed through commercial depuration plants, and sold.

Relatively little effort has been made in this country to enhance stocks of commercially significant crustaceans (lobsters, shrimp, crabs), but some noteworthy attempts do exist. The Commonwealth of Massachusetts, in particular, attempted to enhance its American lobster stocks via release of hatchery-reared juveniles. A small lobster hatchery on Martha's Vineyard was active from 1951 until a recent setback by a major fire, but results from its more than three decades of releases are ambiguous.

Augmentation of wild shrimp stocks with hatchery-reared postlarvae is a well-established practice in some countries, notably Japan. The cost-effectiveness of this type of enhancement is open to serious question, and it has not been attempted to any major degree in the United States. However, some experiments are under way to evaluate the potential for augmenting reproducing shrimp populations following winter kills through the release of wild subadults maintained in captivity over winter. It is believed that these animals would quickly mature and reproduce in the wild, yielding progeny that would subsequently be recruited to the local population at a rate sufficiently high to support some degree of fishing pressure (Sandifer et al., 1991a,b).

ECONOMIC ISSUES

Many factors that directly or indirectly affect costs are likely to determine the future success of marine aquaculture businesses. The major costs affecting the economic feasibility of an aquaculture enterprise are summarized below.

Regulatory-Permitting Costs

The costs of complying with legal and regulatory requirements are substantial in most states. In Maine, for example, a recent survey of salmon farmers indicated that it would take over a year and in excess of $100,000 in fees, research, and legal costs to obtain appropriate permits to begin salmon farming. Furthermore, 70 percent of the respondents expected the permitting and leasing costs to increase (Bettencourt and Anderson, 1990). The permitting process tends to be time consuming and costly, involving a number of federal, state, and in some cases, local agencies.

Capital Costs

If permits can be obtained, start-up capital costs include the following: ponds, tanks, cages, boats, motors, tractors, anchors, moorings, fish transport vehicles, office/warehouse facilities, feed/maintenance shelter, carrying/storage containers, and a variety of culture and handling equipment. A recent study of the southwestern New Brunswick salmonid cage culture industry estimated that a 24-cage site producing 91 metric tons of Atlantic salmon would have total capital costs of approximately $220,000 (1987 U.S. dollars), exclusive of site acquisition costs. Cages accounted for nearly 55 percent of estimated costs (Flander-Good Associates Ltd., 1989).

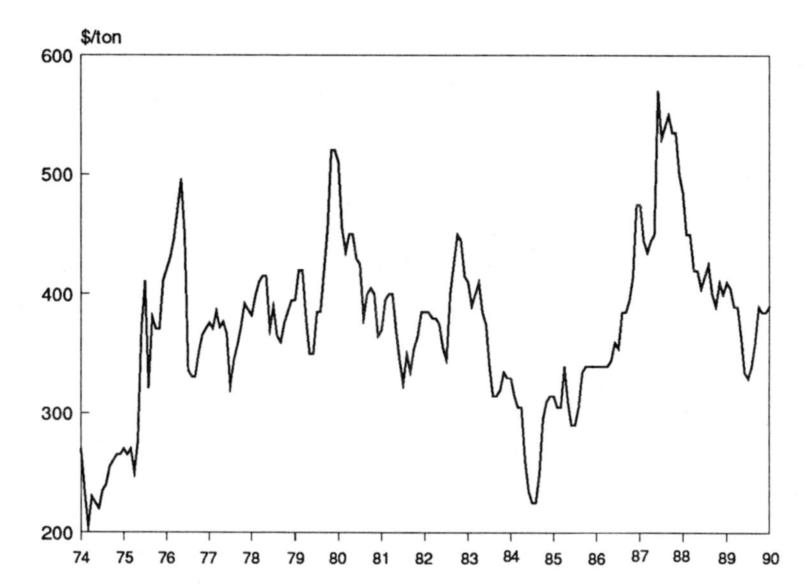

FIGURE 2-9 Fish meal cash prices in Atlanta 1974–1990. SOURCE: *Feedstuffs* (various issues).

Acquisitions of capital and financing for marine aquaculture are major concerns aggravated by uncertainties about the regulatory environment, costs and output prices, performance of the technology, and growth and mortality rates, and by the capital-intensive nature of much of marine aquaculture technology. In addition, the corrosive saltwater environment and the high cost of coastal land tend to increase costs for marine aquaculture in comparison to freshwater aquaculture. With the exception of Farmer's Home Administration programs for shellfish farmers, no guaranteed government loan programs exist for marine aquaculture. In general, U.S. banks are reluctant to finance this fledgling industry, and the availability of venture capital is highly variable. These capital constraints inhibit industry growth and tend to foster the predominance of foreign investors in U.S. marine aquaculture. For example, Canadian and Norwegian interests dominate the salmon aquaculture industry in the United States. Investors from Taiwan recently began operating the largest U.S. shrimp farm in Texas.

Operating Costs

Normal operating costs generally fall into the following categories: smolts/stock, feed, labor, insurance, processing, marketing fees, ice, gasoline and oil, heating and electricity, office expenses, management, and contract maintenance. Flander-Good Associates (1989) estimated operating costs for a 24-cage salmonid site to be on the order of $490,000 per year.

The primary operating cost for most marine finfish operations is feed (usually about 30 percent or more). The price of feed is closely tied to fish meal prices, which correlate closely with soybean meal prices and are highly variable (see Figure 2-9). As aquaculture production increases worldwide, the demand for fish meal and other feed ingredients will increase, possibly driving up feed prices. Developing means by which farmers can achieve better feed conversions and, more important, derive better growth rates per dollar spent on feed is important to the ultimate profitability of aquaculture. Mollusk culture relies on living food, which is usually very expensive compared to formulated diets.

Feed, in addition to being an essential input, can also be a major source of pollution from marine aquaculture. Pollution, in turn, raises the cost of operations through site degradation and the concomitant negative influence on production, and through fines and/or effluent charges, on permit revocations. To be economically successful, not only must feed and feeding practices be cost-effective to yield more growth, they must result in minimal pollution impacts.

Labor is relatively costly in most of the United States compared to many of the countries successfully competing in marine aquaculture. In addition, the aquacultural skill level of the U.S. labor force is relatively low. Increasing costs for both unskilled and technical labor are likely in the future, which will further erode the profit margin for marine aquaculture.

The cost of capital, depreciation, and debt tends to be high in marine aquaculture. A challenge for marine aquaculture is the development of systems (culture systems and auxiliary systems, see Chapter 5) that are effective, yet are not highly capital intensive.

Disease transmission among fish, identification and treatment of disease, and its impact on growth and survival all result in increased costs. The negative image of diseased products also may inhibit market success.

Marketing Factors

Costs of marketing the product to the consumer are often underestimated. For fresh/frozen wild finfish (e.g., salmon), the costs of packaging, processing, transportation, and incidentals usually amount to a markup of at least 100 percent from the ex-vessel price to the primary wholesale price. Secondary seafood wholesalers add another 20 to 25 percent markup to the primary wholesale price; and retailers mark up the price by 25 to 35 percent (USDC, 1990).

Prices are highly variable in the seafood business, as was observed recently in the salmon industry when production temporarily exceeded demand in the world salmon market. Clearly, economic feasibility studies and projections based on constant prices must always be used with caution. Successful marketing of substantial amounts of fish may require larger and more aggressive marketing efforts coordinated among suppliers or through the leadership of a dominant firm. The latter case would tend to emulate the poultry industry model. The strongest selling points for promoting the positive benefits of aquacultural products to consumers are the predictability of product quality and the availability. A key criterion for marine aquaculture R&D is to develop technology that improves these factors to meet the marketing opportunity.

Foreign Competition and Trade

The United States has few barriers to imports of seafood from abroad. Many foreign marine aquaculture industries obtain assistance from their governments through protective trade barriers (i.e., Canada, Europe, and Japan). Additional public support is provided through research and development funds (Norway and Scotland), subsidized transportation (Norway), price supports (Norway), government loan assistance (Canada, Norway), and subsidized market research and development for aquaculture (Canada, Norway, Ireland, Scotland). More discussion of other countries' policies is provided in Appendix A. Unlike U.S. agriculture, U.S. marine aquaculture products are at a disadvantage with foreign competition on the world market.

It is apparent that a number of opportunities exist to reduce the costs of production and marketing through advances in technology, thereby improving the competitiveness of the U.S. marine aquaculture industry. Technology, however, can be effective only if a number of institutional, regulatory, and environmental issues are addressed through the public policy process. Many marine aquaculture technologies and marine species are speculative at this point. Most of the intensive onshore marine systems must be considered speculative, as should offshore systems that are truly exposed to the open ocean environment.

NOTES

[1]Production figures in wet (fresh) weight are rounded throughout this report because of inconsistency or disagreement of more refined estimates in the literature. Weight of mollusks includes that of shells; when only meat weight is given in source material, the assumption is made that meat weight equals 20 percent of total weight.

[2]U.S. per capita consumption figures include domestically cultured oysters, clams, and catfish, but do not include domestically cultured salmon, trout, or other species (U.S. Department of Commerce, National Marine Fisheries Service, Fishery Statistics Division, personal communication, 1991).

[3]*Extensive culture*: Low density in a large area (usually a natural water body), requiring little or no supplementary feeding or environmental management. Production costs are low;

however, lack of control means that production rates are low (<2,000 lbs/acre) and unpredictable.

Intensive culture: Medium to high density, contained in an enclosed area with control of feeding and detrimental factors in the natural environment. Investment costs are high, but there is generally more predictability of outcome, and production rates are higher (<2,000 lbs/acre). Systems are susceptible, however, to stress, disease, and reduced growth from crowding.

REFERENCES

Anderson, J., and J. Wilen. 1986. Implications of private salmon aquaculture on prices, production, and management of salmon resources. American Journal of Agricultural Economics. 68(4):866-879.

Bailey, R. 1988. Third world fisheries: Prospects and problems. World Development 16:751-757.

Bettencourt, S., and J.L. Anderson. 1990. Pen-Reared Salmonid Aquaculture in the Northeastern United States. U.S. Department of Agriculture, Northeast Regional Aquaculture Center Report 100. Kingston, R.I.

Binkowski, F.P., and S.I. Doroshev. 1985. Epilogue: A perspective on sturgeon culture. Pp. 147-152 in North American Sturgeons: Biology and Aquaculture Potential, F.P. Binkowski and S.I. Doroshev, eds., Dr. W. Junk Publishers, Dordrecht.

Boyce, J. 1990. A Comparison of Demand Models for Alaska Salmon, Department of Economics, University of Alaska, Fairbanks, under contract with Fisheries Research and Enhancement Division, Alaska Department of Fish and Game, August. 102 pp.

Chamberlain, G.W. 1991. Status of shrimp farming in Texas. Pp. 36-57 in Shrimp Culture in North America and the Caribbean, P.A. Sandifer, ed. The World Aquaculture Society, Baton Rouge, La.

Chamberlain, G.W., R.J. Miget, and M.G. Haby (compilers). 1990. Red drum aquaculture. Texas A&M Sea Grant College Program No. TAMU-SG-90-603. College Station, Tex.

Chew, K.K., and D. Toba. 1991. Western region aquaculture industry: Situation and outlook report. Western Regional Aquaculture Consortium, University of Washington, Seattle, 23 pp.

Council of Economic Advisors. 1989. Annual Report of the Council of Economic Advisors. Washington, D.C. U.S. Government Printing Office.

Doroshov, S.I. 1985. Biology and culture of sturgeon, Acipenseriformes. Pp. 251-274 in Recent Advances in Aquaculture, Vol. 2. James F. Muir and Ronald J. Roberts, eds. Boulder, Colo.: Westview Press.

Economic Report of the President. 1991. U.S. Government Printing Office, Washington, D.C.

Feedstuffs, The Weekly Newspaper for Agribusiness. Various issues, 1974-1990. Miller Publishing Co., Mannetorka, Minn.

Flander-Good Associates. 1989. Economic Assessment of Salmonid Cage Culture Industry in Southwestern New Brunswick. Fredericton, New Brunswick. 105 pp.

Food and Agriculture Organization (FAO). 1990. FIDI/C:815 Revision 2; as reported in Fish Farming International 17(8):12-13.

Food and Agriculture Organization (FAO). 1991. P. 145 in FAO Yearbook, Fishery Statistics, Catches and Landings, Vol. 68, 1989, Rome.

Gulland, J.A. 1971. The Fish Resources of the Ocean. Surrey: Fishing News Books Ltd. 255 pp.

Hagood, R.W., G.N. Rothwell, M. Swafford, and M. Tosaki. 1981. Preliminary report on the aquaculture development of the dolphin fish, *Coryphaena hippurus* (Linnaeus). Journal of the World Mariculture Society 12(1):135-139.

Harkness, W.J.K., and J.R. Dymond. 1961. The Lake Sturgeon, the History of Its Fishery and Problems of Conservation. Toronto, Ontario Department of Lands and Forests. 121 pp.

Hjul, P. 1973. FAO conference on fishery management and development. Fishing News Internat. (May):20-35.

Holt, G.J. 1992. Experimental studies of feeding of larval red drum, J. World Aquaculture Society. (In press.)

Hopkins, J.S. 1991. Status and history of marine and freshwater shrimp farming in South Carolina and Florida. Pp. 17-35 in Shrimp Culture in North America and the Caribbean, P.A. Sandifer, ed. Baton Rouge, La. The World Aquaculture Society.

Hughes, J.T., J.J. Sullivan, and R. Shleser. 1972. Enhancement of lobster growth. Science 177:1110-1111.

Kraul, S. 1992. Larviculture of the mahimahi, *Coryphaena hippurus* in Hawaii, USA. Journal of the World Aquaculture Society.

Manzi, J.J. 1990. The role of aquaculture in the restoration and enhancement of molluscan fisheries in North America. Pp. 53-56 in Marine Farming and Enhancement, A.K. Sparks, ed. Proceedings of the 15th U.S.-Japan Meeting on Aquaculture. Kyoto, Japan, October 22-23, 1986. NOAA Tech. Report NMFS 85.

Mayo Associates. 1988. An Assessment of Private Salmon Ranching in Oregon. Prepared for the Oregon Coastal Zone Management Association, Inc., Seattle, Washington. 85+ pp.

McCarty, C.E., J.G. Geiger, L.N. Sturmer, B.A. Gregg, and W.P. Rutledge. 1986. Marine finfish culture in Texas: A model for the future. In Fish Culture in Fish Management, R.H. Stroud, ed. American Fisheries Society, Washington, D.C.

McNeil, W.J. 1988. Salmon Production, Management, and Allocation—Biological Economic and Policy Issues, W.J. McNeil, ed. Oregon State University Press.

Meriwether II, F.H., E.D. Scura, and W.Y. Okamura. 1983. Culture of red tilapia in freshwater prawn and brackish water ponds. Proceedings, 1st International Conference on Warm Water Crustacea, Brigham Young University, Laie, Hawaii: 260-267.

Meriwether II, F.H., E.D. Scura, and W.Y. Okamura. 1984. Cage culture of red tilapia in prawn and shrimp ponds. Journal of the World Aquaculture Society 15:254-265.

Naef, F.E. 1971. Pan-size salmon from ocean systems. Sea Grant 70's 2(4):1-2.

Nehlsen, W., J.E. Williams, and J.A. Lichatowich. 1991. Pacific salmon at the crossroads: Stocks at risk from California, Oregon, Idaho, and Washington. Fisheries 16(2):4-21.

Pillay, T.V.R. 1976. The state of aquaculture 1976. In Advances in Aquaculture, T.V.R. Pillay, and W.A. Dill, eds. Surrey: Fishing News Books Ltd.

Pruder, G.D. 1991. Status of shrimp farming in Texas. Pp. 36-57 In Shrimp Culture in North America and the Caribbean, P.A. Sandifer, ed. Baton Rouge, La. The World Aquaculture Society.

Putnam, J.J., and J.E. Allshouse. 1991. Food consumption, prices, and expenditures 1968-1989. Statistical Bulletin No. 825. U.S. Department of Agriculture, Economic Research Service, Washington, D.C.

Ricker, W.E. 1969. Food from the sea. In Resources and Man, P. Cloud, ed. Chicago: Freemand and Company. 290 pp.

Rosenberry, R. 1991. World shrimp farming. Aquaculture Magazine (September/October):60-64.

Royce, W. F. 1989. A history of marine fishery management. Aquatic Sci. 1:27-44.

Ryther, J. H. 1969. Photosynthesis and fish production in the sea. Science 166:72-76.

Sandifer, P.A. 1991. Species with aquaculture potential for the Caribbean. Pp. 30-60 in Status and Potential of Aqualture in the Caribbean, J.A. Hargreaves and D.E. Alston, eds. World Aquaculture Society.

Sandifer, P.A., J.S. Hopkins, A.D. Stokes, and R.A. Smiley. 1988. Experimental pond grow-out of the red drum, *Sciaenops ocellatus*, in South Carolina. Journal of the World Aquaculture Society 19(1):62A (abstract).

Sandifer, P.A., J.S. Hopkins, A.D. Stokes, and G. D. Pruder. 1991a. Technological advances in intensive pond culture of shrimp in the United States. Frontiers of Shrimp Research. Elsevier. New York, N.Y.

Sandifer, P.A., A.D. Stokes, and J.S. Hopkins. 1991b. Further intensification of pond shrimp culture in South Carolina. In Shrimp Culture in North America and the Caribbean, P.A. Sandifer, ed. World Aquaculture Society.

Smith, T.I.J. 1986. Culture of North American sturgeons for fishery enhancement. Proceedings of the 15th U.S.-Japan Meeting on Aquaculture, Kyoto, Japan, October 22-23. NOAA Tech. Report NMFS 85:19-27.

Smith, T.I.J., and W.E. Jenkins. 1991. Development of a shortnose sturgeon, *Acipenser brevironstrum*, stock enhancement program in North America. In Acipenser Sturgeon: Proceedings of the 1st International Bordeaux Symposium 1989, CEMAGREF, Bordeaux, France. Patrick Williot, ed. 520 pp.

Stevens, R.E. 1984. Historical overview of striped bass culture and management. Pp. 1-15 in The Aquaculture of Striped Bass: A Proceedings, Joseph P. McCraren, ed. College Park, Md: University of Maryland. Pub. No. UM-SG-MAO-84-01.

Szyper, J., R. Bourke, and L.D. Conquest. 1984. Growth of juvenile dolphin fish, *Coryphaena hippurus*, on test diets differing in fresh and prepared components. Journal of the World Mariculture Society 15:219-221.

Tilseth, S. 1990. New marine fish species for cold-water farming. Aquaculture 85:235-245.

Urner Barry. Various issues, 1985-1990. Seafood Price: Current. Tom's River, N.J.

U.S. Department of Agriculture (USDA). 1990. Outlook for U.S. Agricultural Exports. Foreign Agricultural Service, Economic Research Service. Washington, D.C.

U.S. Department of Commerce (USDC). 1990. Fisheries of the United States. NOAA/NMFS, Washington, D.C.

Van Olst, J.C., and J.M. Carlberg. 1990. Commercial culture of hybrid striped bass. Aquaculture Magazine 16(1):49-59.

Virginia Sea Grant. 1990. A plan for addressing the restoration of the American oyster industry. Virginia Sea Grant College Program, USG-90-02.

Wilson, J., and D. Fleming. 1989. Economics of the Maine mussel industry. World Aquaculture 20(4):49-55.

Wolniakowski, K., M. Stephenson, and G. Ishikowa. 1987. Tributyltin concentrations and oyster deformations in Coos Bay, Oregon. Pp. 1438-1442 in Oceans '87 Proceedings, Vol. 4, International Organotin Symposium.

3

Policy Issues

Marine aquaculture is subject to a number of policy systems and forces, including direct regulatory regimes imposed by federal, state, and local governments; indirect state and federal economic policies; and the broad array of environmental issues that play an increasingly powerful role in shaping the direction of any activities that affect natural resources and the environment.

THE FEDERAL GOVERNMENT AND MARINE AQUACULTURE

Several major strains in federal policy and programs affect the development of marine aquaculture:

- the federal government's efforts to promote the husbandry of aquatic plants and animals—irrespective of whether these are fresh or saltwater species;
- marine and coastal policies related to regulation and public trust responsibilities in the planning and management of coastal lands and waters;
- interstate and international trade policies that control the movement of cultured species and products both within the United States and internationally;
- economic policies such as taxes, subsidies, and other fiscal levers;
- consumer policies concerned with product quality, safety, and cost;
- environmental regulatory policies that regulate the conduct of marine aquaculture operations, particularly the discharge of effluents and the use of public resources; and

• a potential area for federal policy action—the promotion of stock mitigation or enhancement to preserve and enhance species of importance for food, recreation, or species preservation.

These various strains of federal policy frequently run counter to one another and must eventually be reconciled if marine aquaculture is to develop fully in the United States. Another layer of complexity is added by states and, in some cases, by local governments, which reserve primary jurisdictional authority over marine aquaculture activities in coastal areas and in state waters.

The "Aquaculture" Side of Marine Aquaculture: The Federal Government as Promoter

Government promotion of aquaculture began in the late nineteenth century in response to pressures from sports fishermen. Initially, public involvement took the form of federal support for the artificial propagation of certain sports fish at publicly funded hatcheries run by the U.S. Fish Commission, now the U.S. Fish and Wildlife Service (FWS). The FWS continues to be responsible for technical research and development of freshwater finfish for recreational and commercial fisheries purposes; a network of FWS laboratories is engaged in research on nutrition, disease, genetics, drug restrictions, and environmental effects. Promotion of marine aquaculture research came in the late 1960s and early 1970s, largely as part of a major new federal push for support of ocean science and engineering (Wenk, 1972; Knecht et al., 1988). The National Oceanic and Atmospheric Administration (NOAA), through the National Marine Fisheries Service (NMFS) and the Office of Sea Grant, undertook a major role in aquaculture research on marine, estuarine, and anadromous fisheries, a role that continues. The U.S. Department of Agriculture (USDA) also is responsible for federal R&D activities in aquaculture, through five regional centers, a competitive grants program, and various extension and information services. USDA-funded research addresses freshwater, saltwater, and anadromous species.

The total federal investment in marine aquaculture activities (about $64.8 million per year); is dwarfed in comparison to the level of support (currently about $36 billion per year) and the range of development incentives that traditional agriculture has received for more than a century (Tiddens, 1990). In addition to the difference in scope between aquaculture and traditional agriculture, the disparity is partly explained by the different time frames during which the two enterprises developed.

In the mid-nineteenth century, there was a convergence in the growth of agricultural science and industry at a time when the nation overwhelmingly supported the agricultural enterprise. The United States was, in effect, a nation of farmers, and its political institutions largely reflected this fact. The agricultural growth model of the nineteenth century stands in contrast to the situation that existed when aquaculture development emerged in the mid-twentieth century, "when numerous industries were already established and competing for political clout as well as limited land, coastal sites, and other resources" (Tiddens, 1990). Growth of the political influence of the industry itself has been hindered by internal divisions between producers of saltwater and freshwater species and by crop-specific orientation. These differences have tended to fragment the industry's organizational base.

Federal policy explicitly aimed at the promotion of aquaculture, which began to accelerate in the late 1970s, has in many ways been largely symbolic in character, for the most part, consisting of studies assessing the status of the industry and the dissemination of aquaculture information, rather than the adoption and pursuit of tangible development goals and incentives. Federal aquaculture promotion policy has involved low levels of funding and generally has worked to maintain the diffusion of responsibilities for aquaculture among various federal agencies, an arrangement that tends to create neglect in a bureaucratic system (Tiddens, 1990).

Federal Legislation to Promote Aquaculture

In the late 1970s, a number of reports focused attention on the constraints preventing the development of the aquaculture industry. The *NOAA Aquaculture Plan* (1977) described the problems and potential of aquaculture and aquaculture science, and called for an enhanced federal promotional role in aquaculture akin to that for agriculture, emphasizing that industry could not do the job alone. A 1978 report of the National Research Council (NRC) thoroughly examined the status of aquaculture and found that "constraints on orderly development . . . tend to be political and administrative, rather than scientific and technological. Advances are needed in all areas, but for overall progress, the essential requirements are policy decisions and administrative actions." Recommendations to ameliorate this situation included establishment of a uniform set of aquaculture policies and naming of a lead agency to direct, guide, support, coordinate, and be responsible and accountable for activities among the federal agencies.

Subsequent passage of the National Aquaculture Act of 1980 (P.L. 96-362) provided an important policy statement regarding the national interest in aquaculture. In the act, Congress "declares that aquaculture has the potential for augmenting existing commercial and recreational fisheries and for producing other renewable resources, thereby assisting the United States in meeting its future food needs and contributing to the solution of world resource problems. It is, therefore, in the national interest, and it is the national policy, to encourage the development of aquaculture in the United

States." The act maintains that principal responsibility for the development of aquaculture, however, must rest with the private sector. The secretaries of agriculture, commerce, and the interior were required to prepare a National Aquaculture Development Plan within 18 months of enactment. The purpose of the plan was to identify potential species for commercial development, and to discuss public and private actions and the research necessary to carry out the objectives of the act.

The act also established the Joint Subcommittee on Aquaculture (JSA) in the Federal Coordinating Council on Science, Engineering, and Technology, which was assigned responsibility for increasing the overall effectiveness and productivity of federal aquaculture research, technology transfer, and economic assistance programs. The JSA was composed of representatives from 12 federal agencies, with the chairmanship originally rotating among the secretaries of agriculture, commerce, and the interior. JSA's role was limited to study and assessment, coordination, planning, collection and dissemination of information, and provision of advice to the federal council.

The act also called for the development of a study on capital requirements for aquaculture to document any capital restrictions to aquaculture development, and a study of regulatory constraints to identify and list relevant federal or state constraints on aquaculture development. The results of the latter study were to be used in the development of a Regulatory Constraints Plan to identify steps the federal government could take to remove unnecessary burdensome regulatory barriers. Although the study was completed in 1980 (Aspen, 1981a,b), no follow-up action has been taken to date, and the report itself is difficult to obtain.

In sum, although providing an important statement of policy, the National Aquaculture Act contained few tangible actions to promote development of the industry and focused instead on study, planning, and coordination efforts. Additionally, the act must be viewed in the context of the advent of a fiscally conservative administration intent on privatization and reducing the federal role (Knecht et al., 1988). The act authorized a total funding level of $17 million in fiscal year 1981 (projected to grow to $29 million in 1983), but given growing fiscal constraints characteristic of the early 1980s, no funds were ever appropriated. With no money, understandably, little action took place.

The National Aquaculture Development Plan of 1984 was prepared by the JSA in response to the National Aquaculture Act. In the plan, it is noted that crippling impediments still persist despite the growth in aquaculture production. The report further states that although certain opportunities exist at the federal level, local and state constraints must also be dealt with. The plan, however, again underscored the administration's policy that primary responsibility for the development of commercial aquaculture in the United States rests with the private sector and essentially called for retention of the status quo in federal funding levels. To further the JSA's broad coordination and monitoring of federal aquaculture programs, the plan established three panels: on science, technology, and engineering; economics; and education and technical assistance.

When the report was released, hearings on reauthorization of the National Aquaculture Act were under way. Congressional testimony revealed a split of opinion. On the one hand, representatives of the administration's policy testified in opposition to the act, arguing that existing programs met the needs of industry and that recent successes in aquaculture had shown that further government action was not needed. In contrast to this viewpoint, scientists and others from the aquaculture community testified in favor of reauthorization, pointing out that the act provided an important and necessary policy statement on the national interest in aquaculture, if nothing else. The aquaculture community was strongly in favor of designation of the USDA as lead agency for aquaculture (Tiddens, 1990.)

The 1985 National Aquaculture Improvement Act (P.L. 99-198) reauthorized the 1980 act and enacted two major amendments: (1) USDA was designated as the lead federal agency with respect to the coordination and dissemination of national aquaculture information, and (2) two new studies were commissioned—one on whether existing capture fisheries could be affected adversely by competition from commercial aquacultural enterprises, and a second on the extent and impact of the introduction of exotic species into U.S. waters as a result of aquaculture activities. Funding was authorized at levels lower than under the 1980 act—$1 million dollars for fiscal years 1986, 1987, and 1988 for each of the three main agencies involved in aquaculture: the Departments of Agriculture, Commerce, and the Interior. Again, these funds were never appropriated.

Current Federal Activities to Promote Marine Aquaculture

Notwithstanding the fiscal problems mentioned above, the federal agencies most involved with marine aquaculture—USDA, NOAA, FWS, and the National Science Foundation (NSF)—have continued to play active roles in marine aquaculture research and development efforts (see Appendix C for detailed description). USDA carries out its aquaculture-related programs through five regional centers that fund cooperative research and educational extension programs in aquaculture, the Cooperative Research Service, and the National Agricultural Library. NOAA is involved in aquaculture through the NMFS and the National Sea Grant Program. NMFS involvement includes the operation of salmon hatcheries; research studies on the culture of such species as oysters, salmon, and shrimp; dissemination of aquaculture-related information; and promotion of international markets for U.S. aquaculture products. The National Sea Grant College Program, through its system of research grants to universities and the Marine Advisory Service, has been responsible for the generation of extensive research on biological and technological aspects of marine aquaculture production for marine, estuarine, and Great Lakes species. FWS activities in marine aquaculture are related primarily to the agency's operation of fish hatcheries, fish health centers, fish technology centers, and fishery research centers. Research studies generated at these facilities (e.g., nutrition, disease control, and rearing strategies for Pacific salmon, Atlantic salmon, and striped bass) are directly relevant to marine aquaculture development. The NSF Small Business Innovation Research (SBIR) Program provides funding to small business firms on scientific or engineering issues that could lead to significant public benefit, including research on marine/estuarine and freshwater aquaculture.

The Joint Subcommittee on Aquaculture provides a forum for interagency communication about federal aquaculture activities. It develops and promotes aquaculture through periodic meetings and workshops aimed at fostering coordination of the actions of the federal agencies involved in aquaculture and maintaining communication among affected user groups.

In 1987, for example, the JSA sponsored a National Aquaculture Forum in Davis, California, to establish national goals, identify constraints, and describe opportunities for growth of the industry. Participants in the forum included research and extension scientists, aquaculture industry representatives, and federal government administrators. A number of action strategies were identified to enhance the growth of the aquaculture industry in the following areas: marketing, production efficiency, processing and product development, industry representation at regional and national levels, awareness of global aquaculture technology, integration of aquaculture with traditional agriculture, expanding the role of the private sector in fish enhancement programs, promoting state and federal research and development, streamlining permitting processes, obtaining approval for therapeutic compounds, and information systems (JSA, 1990).

In the past two years the JSA has addressed topics such as reports from the USDA Regional Aquaculture Centers; the formation of the National Aquaculture Association; a memorandum of understanding between the USDA and the FWS on fish health management; protective statutes (such as the Lacey Act); issues of research and technology transfer; formation of an interagency working group on effluents from aquaculture operations; and activities of the National Aquaculture Information Center (NAIC) of the National Agricultural Library (JSA, 1991).

Summary: Status of Federal Promotional Policy

Despite legislative efforts to rationalize and coordinate the federal government's promotion of aquaculture, the various federal agencies that deal directly with marine aquaculture by and large continue to pursue their traditional roles vis-à-vis marine aquaculture on separate tracks. Although the JSA provides a forum for discussion and recommendation of strategies for improving the outlook for aquaculture, several institutional problems hinder its effectiveness: (1) the lack of authority over the programs of its many federal agencies, (2) the predominance of freshwater aquaculture concerns on the JSA agenda; and (3) location of the JSA under the umbrella of the President's Office of Science and Technology Policy, which subjects it to inevitable executive branch policy swings.

In summary, although individuals in federal agencies have made personal efforts to fulfill the mandate of the stated national policy to encourage and stimulate the expansion of aquaculture, the federal government's "promotional" role regarding marine aquaculture has been confined largely to general policy statements of support, the conduct of repeated studies on the obstacles facing aquaculture, and the formation of interagency mechanisms that, while promoting the important goal of coordination, lack any substantial power and authority. This is in sharp contrast to the agricultural model and to the experience in other nations (see Appendix A for an analysis and review of other countries' aquaculture policies). For marine aquaculture to succeed in the United States, a more active and forceful federal role will be needed—one that employs a wider range of incentives for aquaculture development akin to those used in the development of agriculture and that centralizes authority (and corresponding resources) to support the promotional role in the lead federal agency, the USDA.

In order to implement the intent of the legislation that has been enacted over the past decade to encourage the development of aquaculture, active congressional oversight is necessary. A mechanism for exercising such oversight is a congressional committee or subcommittee. Such a committee would be responsible for ensuring that executive agencies coordinate their aquaculture-related activities to achieve the maximum efficiency in the use of limited resources and that sufficient funds are appropriated to carry out the legislative mandate of the National Aquaculture Act and National Aquaculture Improvement Act.

The "Marine" Side of Marine Aquaculture: The Federal Role in Planning and Regulating Coastal Commons

Marine development operations take place in or near coastal lands and waters, a special realm over which the federal government has important public trust responsibilities. The coastal ocean has been traditionally a public space—open to all to use and enjoy—where resources have been viewed as common property. Moreover, a high degree of interconnection exists between ocean resources and marine processes; users of ocean resources are ultimately interconnected and are inevitably affected by changes in the overall health of the ocean. The coastal ocean is also an area in which all levels of government—federal, state, and local—play a role, which complicates the management picture. The land side of the coastal zone is also a special area—a highly limited but unusually valuable place where land and sea meet. It is a focus for recreation and enjoyment, commerce, and industry, and it is valued also as an area of unique ecological significance. The use of shore lands affects coastal waters, and conversely, the forces of the sea shape shore lands and their uses.

Because of the special and largely public character of the coastal zone (both land and sea), the federal government performs a variety of functions in this area. First, there is a *public trust and regulatory function* to ensure that ocean and coastal resources are protected for both current and future generations. A *conflict resolution function* exists to mediate competing claims of the many users of ocean and coastal resources and space. Further, there is a *proprietary function* to obtain a fair return to the public for the rent of submerged lands to private interests for exploitation and profit-making purposes (Knecht, 1986). The last responsibility means that marine aquaculture operates under conditions in contrast to the practice of freshwater aquaculture, which takes place largely on private property where the rights to the aquaculture products are clear and the governmental regulatory role is limited. Marine aquaculture operations must compete with many other users for access to limited, valuable, and generally public coastal lands and waters, where the rights to the product are undefined and an assortment of government agencies wields extensive regulatory power.

Ocean and Coastal Zone Legislation

The U.S. Congress has enacted about a dozen laws aimed at protecting and managing the ocean and coastal zone in response to the widespread concern over the health of the oceans and coastal areas that emerged on the public agenda in the 1970s. Among the most important laws promulgated during this period are the Coastal Zone Management Act of 1972 (P.L. 92-583); the Clean Water Act of 1972 (P.L. 95-217); the Marine Protection, Research, and Sanctuaries Act of 1972 (P.L. 92-532); the Marine Mammal Protection Act of 1972 (P.L. 92-532); the Endangered Species Act of 1973 (P.L. 93-205); the Magnuson Fishery Conservation and Management Act of 1976 (P.L. 94-265); and the Outer Continental Shelf Lands Act Amendments of 1978 (P.L. 95-372).

Perhaps the major characteristic of this body of law is that, with the exception of the Coastal Zone Management Act, most federal laws dealing with the ocean and coastal zone tend to be single purpose in nature, each statute promoting a particular aspect of the marine and coastal environment (e.g., environmental protection, fishing, oil and gas development). Inconsistencies among federal laws and programs thus are difficult to resolve (e.g., the conflict between environmental protection and promotion of leasing for oil and gas resource development). Conflicts among multiple uses of the coastal zone and ocean have escalated as this body of law has been implemented in the 1980s and 1990s. Under the current federal regulatory framework, it is difficult to solve such conflicts or to plan the development of various ocean and coastal uses in specific areas (Cicin-Sain, 1982; Cicin-Sain and Knecht, 1985). The problems that arise from the lack of a clear policy are evidenced by a recent suit by a number of New England environmental organizations against the U.S. Army Corps of Engineers for issuance of a permit to American Norwegian Fish Farm, Inc., to establish a 47-square-mile offshore salmon farm 37 miles east of Cape Ann, Massachusetts. Questions that have arisen from the suit include whether or not such an enterprise is the best use of public waters, whether the fish farm should be charged a lease fee, and whether an environmental impact statement should be required. At present, none of these questions are addressed in a management policy for federal waters (National Fisherman, 1991).

Assessment of Ocean and Coastal Policies

Marine aquaculture has a relatively weak base in the conflict over the use of coastal ocean resources and space, both in the regulatory framework and in the political arena, compared to more established groups promoting other uses of the ocean and coastal environment (e.g., fisheries, oil and gas).

Politically, marine aquaculture is less well organized and has fewer resources than more established groups promoting other uses of the ocean and coastal environment (such as fishing and oil development). As a result, marine aquaculture is often ignored and, consequently, loses out to more influential interests in public deliberations over the use of specific ocean and coastal areas. Contributing to the political ineffectiveness is the lack of public support or even recognition of marine aquaculture as a beneficial, food-generating enterprise. In some cases, such as controversies over salmon pens in the state of Washington, the limited public knowledge that does exist about aquaculture tends to be associated with the pollution aspects of aquaculture operations and is more negative than positive (Chasan, 1990).

The lack of a positive regulatory and political basis for aquaculture has been exacerbated in recent years by the resurgence of the environmental movement (after a decline in the late 1970s). This resurgence has been accompanied by a focus of attention on environmental issues of coastal water quality and wetlands protection, two areas that significantly affect marine aquaculture operations. In response to public pressures, new federal initiatives by the Environmental Protection Agency (EPA) and by NOAA have been undertaken to improve coastal water quality and to prevent the further loss and degradation of wetlands (the "no-net-loss" policy). This trend is likely to lead to more stringent scrutiny of aquaculture operations. (See Chapter 4 for further discussion of environmental issues and marine aquaculture.)

International and Interstate Trade Policies

International and interstate trade policies and incentives directly impact the profitability and competitiveness of the U.S. aquaculture industry.

International Trade Issues

The United States imposes minimal restrictions on the importation of seafood. Most imports that compete with aquaculture products enter with no or minimal duties. In contrast, substantial tariffs or nontariff barriers are often imposed for U.S. seafood exported to Canada and the European Community even though many of the U.S. agriculture industries are protected to some degree through import quotas and tariff barriers. Partially protected industries include dairy products, sugar, and many fruits and vegetables. The free trade position in the United States with regard to most seafood gives foreign competition access to valuable markets and a potentially unfair advantage in the market through subsidies or other incentives.

A suit has recently been brought against the Norwegian fresh and chilled farmed Atlantic salmon industry. The U.S. producers of Atlantic salmon have alleged that the Norwegians are able to sell salmon at less than fair value because the Norwegian producers have received considerable subsidies from their government. The U.S. International Trade Commission found that the Norwegian government subsidized its salmon industry through a variety of regional development loans and grants, regional capital tax incentives, federal payroll taxes, and advanced depreciation on assets. In addition, the commission ruled that the Norwegians have dumped Atlantic salmon by selling at less than the cost of production. The result of the investigation led to the first significant tariffs to protect the marine aquaculture industry from "unfair" competition (Helm, 1989).

However, such protection may have come too late for many U.S.-owned companies. Ocean Products, Inc., in Eastport, Maine, which spearheaded the case, recently met with severe financial difficulty and is now 100 percent Canadian owned. In addition, most salmon farming in the United States is currently under some degree of foreign ownership.

The aquaculture industry also may qualify for assistance through the Export Enhancement Program authorized under the 1985 Food Securities Act and other export-oriented programs. These programs assist food-producing groups in developing and maintaining market shares abroad. To date, the marine aquaculture industry has not taken advantage of these possible opportunities.

The aquaculture industry in the United States needs to become more informed about trade laws and programs that may be used to its benefit. In addition, U.S. trade policies need to address issues of foreign practices that are illegal in the United States, such as the use of certain drugs and chemicals. Use of these substances may lower production costs for foreign competitors and may also pose a health risk to U.S. consumers.

Interstate Trade Issues

Several barriers hamper interstate trade within the United States. One example is the Lacey Act and the Lacey Act Amendments of 1981 (Title 16, U.S.C. 3371), which regulate, among other activities, the movement of live fish between states. The Lacey Act has two principal purposes: (1) prohibition of commerce in unlawfully taken wildlife, and (2) prevention of the introduction of injurious species of wildlife in the United States. The act makes an important contribution to environmental preservation and protection of indigenous species; however, its linkage with state laws—which are not uniform and are perceived to be outdated and unreasonable in some cases—has created a mechanism that often discourages commercial aquaculture. In particular, the designation of state borders as geographic control points can result in an arbitrary restraint of trade; for example, the transport of live fish a few miles from Rhode Island to Connecticut within the same ecological zone may violate the Lacey Act. This situation imposes additional costs on aquaculturists, especially in New England, and frequently interferes with normal marketing practices (USDA/USDI, 1990).

A related area of regulation within the states is associated with fisheries management. When fish size or harvest constraints are imposed on the wild fishery without regard for aquaculture, aquaculture activity may be eliminated or constrained without reason. For example, in an effort to protect the wild population of striped bass on the East Coast, some states imposed regulations that make possession or sale of striped bass illegal. This restriction, although aimed at protecting the genetic diversity of wild populations, has prevented the development of striped bass aquaculture enterprises. Alternative approaches might have protected the wild fisheries without negative effects on aquaculture. The restriction also limited interstate trade because it was illegal, in some cases, to transport striped bass from states where the product was legally grown to states where it was legal to sell. A more reasonable set of restrictions would reflect valid ecological and biological considerations, rather than political jurisdictions.

A joint USDA/USDI work group examining protective statutes relating to aquaculture reached the following conclusions regarding the Lacey Act (USDA/USDI, 1990): (1) policy is needed to define and elaborate the dual role of fish as livestock and as a public resource, and (2) uniform and consistent state and federal laws and regulations regarding aquaculture and aquaculture species are necessary.

Economic Policies

Federal programs that relate to financing of marine aquaculturists are very limited, although some direct economic assistance to businesses does occur through public programs at both federal and state levels. The USDA's Farmers Home Administration is the government's largest provider of investment and operating capital to aquaculture. The majority of this support is to the catfish and oyster industries. Other agencies that could provide assistance to aquaculture include the Small Business Administration, the Economic Development Agency, and the Farm Credit System. At present, a federal policy specifically directed to provide aquaculture credit does not exist. Nor are there any specifically formulated tax or agency incentives for marine aquaculture at this time, although the prospect does exist for significant incentives at the state and local government levels.

Shellfish culturists generally are linked closely to the shore-sea interface. Zoning and water quality regulations are important factors affecting their operations. Because of the evolving nature of marine aquaculture technology, state and local governments may control more of the regulatory variables than the federal government. Cumbersome, time-consuming requirements of a multilayered permitting process, complex regulations, and a general indifference on the part of most local government agencies have resulted in substantial costs to the industry, especially to those attempting to begin operations. In contrast, the U.S. agriculture sector has a vast array of programs for farms, including deficiency payments, nonrecourse loans, emergency compensation, paid diversion, export enhancement programs, disaster payments, marketing loans, and many others.

Consumer Policies

Development of policies and regulations that protect consumer interests and welfare must not be overlooked as the aquaculture industry grows. Ensuring seafood quality, safety, and wholesomeness is important not only for the consumer but also for the long-run stability of the industry. Any case of illness or death resulting from consumption of seafood, whether wild or farmed, can have disastrous economic repercussions, not only for the producers and handlers of the species involved but also for the seafood, marine, and agriculture industries in general.

Two important areas of public policy that impact the seafood consumer are informative labeling and product safety. Buyers need objective information regarding a product's supplier, its ingredients, and the nutritional value if they are to make informed decisions that in turn lead to selection of high-quality products, consumer satisfaction, and repeat purchases. Brand names, advertising, and product labels have been used by food sellers since the nineteenth century to ensure repeated sales to satisfied customers. Yet, more than 100 years later, virtually none of these practices have found their way into the fresh seafood industry, with the exception of the catfish market. Instead, consumers tend to rely on the retailer for quality cues when purchasing fresh seafood. Accurate identification of seafood species is also important consumer information. Approved trade names for fish species exist; however, enforcement is limited and seafood (especially high-value seafood) is sometimes mislabeled fraudulently to obtain a better market price. In addition, there is currently no policy having to do with whether seafood should be labeled as farmed or wild, and whether labels should include other informative items such as country of origin, areas of harvest, date of harvest, and related documentation.

Many consumers desire more detailed information regarding quality and safety characteristics. A recent report (IOM, 1991) identifies the major risk from seafood to be the consumption of raw bivalves. Other risks include consumption of seafood taken from polluted waters. The major sources of pollution cited are human waste and chemical runoff. Chemical and microbial contamination in culture water and the use of therapeutic drugs are growing concerns with aquaculture products (IOM, 1991). These issues are discussed in detail in Chapter 4.

The Food and Drug Administration (FDA) is currently the primary authority in setting and enforcing regulatory limits having to do with seafood safety. NMFS operates a voluntary fee-for-service seafood inspection program, and EPA sets limits on pesticides in seafood. State agencies also play an active role in the control of seafood safety. Federal programs actually inspect only a small percentage of domestic and imported seafood.

In 1990, legislation to mandate seafood inspection received significant attention from the Congress. The Senate passed the Fish Safety Act of 1990, which gave primary authority to the USDA. Three bills were proposed in the House of Representatives, but no consensus was reached. Although mandatory inspection legislation has not yet been passed, momentum still exists for its enactment in the near future.

In late 1990, Congress passed the Nutrition Labeling and Education Act of 1990, which requires major changes in nutritional and ingredient labeling, and places restrictions on health messages in advertising. As part of this legislation, retailers are mandated to provide nutritional information on the top 20 varieties of raw fish consumed most frequently. This requirement may afford sellers of aquaculture products an opportunity to provide information directly to consumers about the many positive nutritional aspects of seafood and possibly to initiate branding programs.

Seafood inspection programs may be of greater benefit to sellers of farmed products than to sellers of fisheries products because aquaculture firms are likely to have greater control over factors such as water quality, drug residues, harvest, and storage conditions than handlers of wild seafood products.

Shellfish growers, in particular, may benefit from better controls on water for growing shellfish and on shellfish inspection. Aquaculturists, however, may use a variety of therapeutics, chemicals, antifoulants, pesticides, hormones, and related substances to control problems such as disease and parasites and, therefore, may have greater difficulty than the wild fisheries in complying with certain aspects of seafood safety regulations. FDA is responsible for monitoring and regulating the use of drugs in aquaculture.

Nutritional labeling will increase the consumers' knowledge of the nutritional benefits of seafood relative to other protein sources and could enhance the market and opportunities for aquacultural products.

FISHERIES ENHANCEMENT CONCERNS AND POLICY ISSUES

Aquaculture is an important part of fisheries enhancement. The uniqueness of this role necessitates special consideration and policy appropriate to stock enhancement. The role of the private sector in all forms of stock enhancement, including ocean ranching, in public waters is poorly defined.

The question of who should produce fish for enhancement or mitigation purposes is an important policy issue. A considerable amount of freshwater enhancement activity is based on the contracting or purchasing of hatchery-reared stock from private growers. There is no obvious reason why private growers should not be used as primary suppliers for public stock enhancement efforts in marine waters as well, under appropriate contract conditions. A policy of increased use of private hatcheries for enhancement of marine species would stimulate private aquaculture. Approaches to consider include (1) species that are not used currently for enhancement and (2) public/private joint ventures to evaluate, improve, and perhaps expand marine fishery mitigation or enhancement efforts.

In the late 1960s and early 1970s a number of enthusiasts began to visualize the possibilities of private ocean ranching as a commercial undertaking. The availability of salmon propagation technology developed by state and federal agencies and a common perception that private industry could "do it better" encouraged investors in nearly every temperate country to consider salmon culture. In response to increased activity in this area, the four Pacific Coast states each examined the issue: Should private salmon ranching be permitted and if so, under what conditions? Each state adopted a different course:

- Alaska put fisherman-owned, private nonprofit cooperatives in control. A number of facilities have been built from harvest taxes. Their program appears to be successful and expanding. However, for-profit operations are prohibited.
- California passed a law allowing private ocean ranching and then issued only one permit. The returns are minimal and private ocean ranching is not considered a success.
- Washington refused the concept, except for several small, nonprofit efforts. Attempts to change the legislative mind have failed, and salmon pens and tank farms appear to be the form of the future, siting issues notwithstanding.
- Oregon, in the early 1970s, passed laws to make private ocean ranching possible and became America's testing grounds for the concept. In four years (1974–1977) 12 permits were issued and significant construction was undertaken. In 1977, Crown Zellerbach applied for a permit and the same issues that defined the legislative debates were reargued. At that time, however, the courts made the decision and the Crown Zellerbach application was rejected. It is generally considered that the reversal resulted in the current moratorium on new or expanded permits.

Pressures for increased marine fishery enhancement are likely to grow substantially as fishing pressures continue to increase and wild catches decline. At the same time, controversies over the practices and procedures for stock enhancement will grow. In light of the likely expansion of marine stock enhancement efforts, the nation needs a comprehensive policy to guide its actions. Such a policy should provide for the following:

- A careful, unbiased evaluation of past and present stock enhancement programs and practices should be undertaken with regard to their efficacy, cost-effectiveness, environmental problems, and potential payoffs. This evaluation should provide the basis for development of guidelines for sounder, more effective, and less potentially damaging stock enhancement programs in the future.
- Public agencies should ensure that common-property fish stocks are maintained in a healthy condition, that genetic resources are conserved and biodiversity is maintained, and that threatened or endangered species are protected and, where possible, rescued.
- Public agencies should promote the participation and increase the role

of the private sector and the free market system in meeting the nation's needs for fish and shellfish for stock enhancement efforts.

• The private sector needs to meet the specifications required by the public sector for fish and shellfish to be released into public marine waters so that potentially negative environmental effects are minimized or eliminated.

Studies conducted in at least two states (California and Colorado) concluded that savings would occur if some part of their enhancement programs were based on direct purchases of fish from private growers (Mayo Associates, 1988b). Such governmental purchases would also encourage the development of private aquaculture, thus stimulating economic growth. From the technical standpoint, greater involvement of the private sector in the production of fish and shellfish for enhancement would increase communication and technology exchange opportunities between public and private hatcheries. This "cross-fertilization" likely would lead to improvements in technology in both public and private sectors.

THE STATES AND MARINE AQUACULTURE

The federal role in the promotion and regulation of marine aquaculture is circumscribed; in fact, the majority of laws and regulations that specifically authorize, permit, or control aquaculture operations are found at the state level. Significant differences exist among the states regarding marine aquaculture. Marine aquaculture is practiced in the coastal states with varying degrees of acceptance, hostility, regulation, and indifference, depending on available resources, social and cultural traditions, local politics, and the state's economic condition (Davies, 1990). Although some states have designated agencies and formulated plans to promote and assist aquaculture, others do not recognize the industry through any formal structure. In some states, the management of marine aquaculture is vested in the agriculture department; in others, management authority is lodged in the marine resources agencies. Some states, such as Hawaii, have made significant investments in the development of aquaculture through such methods as the designation of a lead agency, the development and adoption of a statewide aquaculture plan, and the creation of marine aquaculture parks that promote development on prepermitted sites (DeVoe and Mount, 1989).

The available literature synthesizing and comparing state experiences with marine aquaculture is scant. Some case studies of the experience of individual states are available (e.g., Tiddens, 1990) as are some studies comparing specific aspects of marine aquaculture regulation (DeVoe and Mount, 1989). A comprehensive report contracted by the USDA in 1980 (Aspen, 1981a,b) examined in detail the aquaculture laws and regulations of 8 states and canvassed the regulatory scene in less detail in 24 other states. Problems with this study, however, include the facts that the report is essentially inaccessible (one noncirculating copy is available in the National Agricultural Library); the study is more than 10 years old, and much has changed at the state level in the last decade; and the focus is not explicitly on marine aquaculture.

The findings of the Aspen report, although dated, remain instructive and have been echoed, in part, in expert testimony received during the course of this investigation. The Aspen study found that regulation of marine aquaculture operations is more stringent than freshwater aquaculture operations and that the most difficult, time-consuming, and costly hurdles were presented by land and water use regulations, such as water appropriation, stream alteration, coastal zone land use, wetlands permits, and special management area permits. No clear regional patterns explaining variations in the severity of compliance burdens were found. Some variations seemed to be related to the level of public awareness of the commercial use of natural resources. The study underscored the point that the local process in permitting of aquaculture can sometimes be more difficult and more time consuming than the state and federal permitting processes.

Other problem areas also were identified in the Aspen report. Property rights in intertidal areas under state jurisdiction, including submerged land and vertical column leases, need clarification, particularly where traditional fishing interests and aquaculturists are competing for use of the same resources (for a detailed examination of this issue, see Davies, 1990). The effects of federal and state coastal wetlands laws on the development of aquaculture were identified as a major emerging issue. Potential impacts of pollutants (including pesticides, radioactive wastes, toxic substances, and acid rain) on environments suitable for aquaculture were highlighted as needing further examination to improve understanding of long-term effects.

The Aspen report found great variations among the states with regard to promotion of aquaculture. Hawaii was found to be the most aggressive and successful state in its support of aquaculture. In contrast, the report noted that Florida had a rigid regulatory stance that tended to discourage aquaculture development. Efforts to streamline the permit process and to create one-stop licensing of aquaculture operations, the report noted, had proved particularly effective in Hawaii, Oregon, and Maine. The variety of options available for permit streamlining highlighted in the report include the following: joint applications for state and federal permits, one-stop permitting procedures, identification of a lead agency to guide applications through interagency comment and review procedures, preapplication consultation with applicants to weed out unacceptable proposals, and aquaculture planning office investigations to provide technical assistance on pond design, disease prevention, methods of preventing predation, and escape of nonnative species into state waters (Aspen, 1981a,b).

Assessment of State Policies

A recent informal survey of industry and public officials in the East and Gulf Coast regions conducted by Richard DeVoe (South Carolina Sea Grant Consortium) identified the following major problems and use conflicts constraining marine aquaculture: recreation (fishing, boating); commercial fishing; limited space (low number of adequate site locations); development (industrial, residential, land use issues); environmental/resource concerns (water availability, pollution, wetland impacts, nonindigenous species); aesthetics; lack of a lead agency; and theft and vandalism (of organisms and facilities) (DeVoe, 1990).

In a number of cases, conflicts between marine aquaculture and other users of the ocean and coastal zone, in fact, have worsened in the past decade. In Washington State, for example, the further development of salmon aquaculture has been hindered significantly by the concerted opposition of property owners, fishermen, and environmental groups who cite such concerns as water pollution from fish excrement, nutrient loading from feed, introduction of disease and antibiotics, as well as loss of fishing grounds and obstruction of view (D.E. Ortman, Friends of the Earth, personal communication, 1990).

Marine aquaculture has also been controversial in Alaska where the state has, in effect, imposed a ban on private for-profit salmon aquaculture largely because of competition with the commercial salmon industry (Hetrick, 1991). In Oregon—initially hailed as the U.S. testing ground for the development of private ocean ranching—the development of this aspect of the marine aquaculture industry has been hampered severely by opposition from commercial fishing and other groups that express fears about the potential genetic effects of ocean ranching operations on the genetic make-up of the natural population (Mayo Assoc., 1988a).

To cite another example, a recent hotly debated issue in South Carolina concerned the importation of nonindigenous marine species for aquaculture purposes. Importation of nonnative penaeid shrimp has raised concerns over the possibility of transfer of the infectious hypodermal and hematopoietic necrosis (IHHN) virus to native populations if animals escape from their culture environments (Manci, 1990).

A proposal for a large marine aquaculture project offshore in Massachusetts recently was halted through a legal suit brought by the Conservation Law Foundation on behalf of several fishing and environmental groups. The suit cited, among other reasons, concerns about possible impacts on navigational rights and the fact that a large area of the public ocean would be privatized without sufficient safeguards or compensation for the public (Conservation Law Foundation, 1990).

On the other hand, the past decade also has seen significant efforts by states to promote marine development. A number of states have prepared and adopted some type of marine aquaculture development plans, named lead aquaculture agencies, or taken other promotional actions, as the following examples indicate.

Hawaii was the first state to carry out comprehensive aquaculture resource planning and development through an Aquaculture Development Plan (presented to the state legislature in 1979) and subsequent creation of the Aquaculture Development Program as part of the state's Department of Land and Natural Resources (Corbin and Young, 1988). The initial planning effort assessed opportunities for and constraints on the development of the industry and identified prime sites for potential marine aquaculture development on the basis of such factors as elevation, slope, soils, surface and groundwater resources, and existing zoning regulations. Characteristics such as ownership, attendant infrastructure, and permit requirements also were included in the site assessment process.

The Hawaii Aquaculture Development Program operates on the concept of the "aquaculture development niche," an approach based on the view that "successful projects require a mixture of both technical and non-technical inputs, that are not all species-related, but include consideration of suitable and available resources and the broad needs of recipient communities" (Corbin and Young, 1988, p. 632). Factors such as markets, support services, sociology, culture, politics, and public policy are included in the assessment of site development potential. Strategic development plans are prepared for each niche opportunity. As part of its strategic planning, Hawaii is experimenting with the establishment of commercial aquaculture parks—areas that have received all the necessary permits to start up and conduct aquaculture operations and that co-locate production enterprises and support services. This comprehensive approach to resource planning has been instrumental in the development of the aquaculture industry in the state and has contributed to the fact that aquaculture is one of Hawaii's fastest-growing ocean industries (McDonald and Deese, 1988).

The need for more effective policy for marine aquaculture development in Maine was documented in a 1987 report (Maine, 1987). In 1989, the Maine State Planning Office and the Department of Marine Resources issued a follow-up report that addressed development requirements for aquaculture, education, state agency regulations, water quality, and consideration of the needs of aquaculture and traditional fisheries (Maine, 1989).

Following the recommendations of the Delaware Aquaculture Task Force, in 1990, the Delaware legislature enacted the Delaware Aquaculture Act. This act recognized aquaculture as a part of agriculture, designated the

Department of Agriculture as coordinating agency for aquaculture, and established a Delaware Aquaculture Advisory Council to enhance and promote aquaculture, identify methods for simplifying the regulatory process, and examine research and educational needs. In support of these efforts, in 1991 the University of Delaware's Graduate College of Marine Studies established an Aquaculture Resource Center to carry out a range of scientific studies and extension services to facilitate the development of aquaculture in the state.

In North Carolina, the Aquaculture Development Act of 1989 designated aquaculture as agriculture, named the North Carolina Department of Agriculture as lead agency, and created the Aquaculture Advisory Board to review and recommend policies, laws, and regulations and to coordinate the activities of state agencies. Two companion pieces of legislation were also enacted—the Water Column Leases for Aquaculture (specifying conditions under which shellfish aquaculture leases could be approved) and a law providing for fines and penalties for damage to aquaculture operations.

In 1985 the legislature in South Carolina created a Joint Legislative Committee on Aquaculture, with a commitment to develop state policy and initiate state programs for aquaculture development. A strategic plan for aquaculture development in South Carolina issued in 1989, identified opportunities and constraints and offered forty-one recommendations that, if implemented, are expected to accelerate the growth of the industry. A 1988 legislative action established an Aquaculture Permit Assistance Office and proclaimed aquaculture as "an important form of both fisheries and agriculture," whereby the Department of Agriculture is responsible for the coordination of promotion and marketing programs and permit assistance, and the South Carolina Wildlife and Marine Resources Department is to handle aquaculture law enforcement and coordination of R&D.

The Georgia legislature designated aquaculture as a form of agriculture in 1987 and, in 1989, created an aquaculture commission and called for the development of a state aquaculture plan.

In 1984, through the Aquaculture Policy Act, the Florida legislature designated the Department of Agriculture and Consumer Affairs as the lead agency for the preparation of a state aquaculture plan (which was completed in 1985), created an Aquaculture Review Council in the Florida Department of Agriculture and Consumer Affairs, and set up an interagency coordinating board to make recommendations to the council.

Although California has no specific aquaculture plan, an aquaculture development section was created in the Department of Fish and Game through legislation passed in 1982. Also created at that time were the Aquaculture Industry Advisory Committee and an interagency committee for aquaculture development. The latter prepared a guide to California's aquaculture industry in 1988 (California, 1988). The California Aquaculture Association, a producer-based organization, has played an active role in the development of the industry and, in 1990, issued a strategic plan for enhancement of the industry in that state (California Aquaculture Association, 1990).

As evidenced in the preceding summary of state aquaculture-related actions, a number of states have taken important steps to enhance the development of aquaculture. However, it is not at all clear whether such actions are making a tangible difference to the economic success of the industry. There are few data available on such important questions as, To what extent are statewide aquaculture plans being implemented, and with what effects? Are the "lead" agencies "leading," with tangible actions to promote the welfare of the industry? What is the level of effectiveness of the variety of interagency committees on aquaculture that have been established? It is clear that a state role is crucial for marine aquaculture and that a better understanding is needed of the states' collective experience in promoting and regulating marine aquaculture, the range of methods used, and the extent of success of various policy approaches.

MANAGEMENT FRAMEWORK FOR MARINE AQUACULTURE

Coastal Zone Management Act

It is apparent that the states have substantial authority to plan for and manage various uses in the coastal zone, both on the land side and on the water side, up to current limits of state authority (3 miles offshore, in most cases). Hence, the integration of marine aquaculture planning and management into the broader framework of coastal planning and management depends to a large degree on state coastal management entities.

States receive assistance for coastal planning and management from the federal government in the form of grants-in-aid given to states under the Coastal Zone Management Act of 1972 (CZMA), as amended. Under the provisions of CZMA and subsequent amendments, states may initially receive federal grants (on a matching basis) for the preparation of coastal management plans or program development grants [section 305]. Once the states prepare coastal plans that meet national standards set forth in the act and obtain federal approval, they become eligible for other federal grants [section 306] to carry out the provisions of the state's coastal plan. Moreover, through the "consistency" provision of the CZMA [section 307], "each federal agency activity within or outside the coastal zone that affects any land or water use or natural resource of the coastal zone shall be carried out in a manner which is consistent to the maximum extent practicable with the enforceable policies of approved State management programs." Thus, the federal government is in a position to influence the nature and conduct of coastal zone management in the 33 coastal states and territories, through the twin incentives of providing funds for program planning and implementation, and of granting the power of "consistency" review to states with federally approved coastal plans. NOAA's Office of Coastal Resources Management is charged with implementation of the Coastal Zone Management Act.

Several approaches can be suggested for better integrating marine aquaculture into the existing coastal zone management process through amendments to CZMA. The first option would be to include a stronger reference to the importance of marine aquaculture in the initial part of the act than now exists. Currently only one reference is made to marine aquaculture in these sections. Section 303(2)(I) states that the Congress finds and declares that it is the national policy "to encourage and assist the states to exercise effectively their responsibilities in the coastal zone through the development and implementation of management programs . . . which programs should at least provide for . . . assistance to support comprehensive planning, conservation, and management of living marine resources, including planning for the siting of pollution control and aquaculture facilities within the coastal zone . . . ."

To add strength to this provision, a separate provision could be included detailing the potential importance of marine aquaculture to the nation in terms of both food production and reduction of the negative balance of payments. Alternatively, one could add marine aquaculture to the list of "coastal-dependent uses" that are identified for priority consideration in coastal planning and management [section 303(1)(2)(C)].

Another option to better integrate marine aquaculture into the coastal management framework would be to explicitly include it as an activity eligible for "Coastal Zone Enhancement Grants" [section 6210]. Enacted as part of the 1990 amendments to the CZMA, this section establishes a program, beginning in fiscal year 1991, to encourage continual improvements in state management programs in one or more of eight identified areas that presently include coastal wetlands management and protection; natural hazards management (including potential sea level and Great Lakes level rise); public access improvements; reduction of marine debris; assessment of cumulative and secondary impacts of coastal growth and development; special area management planning; ocean resource planning; and siting of coastal energy and government facilities. A new provision on marine aquaculture could be added to encourage the preparation, in conjunction with the relevant state fisheries and aquaculture agencies, of state marine aquaculture plans that would assess the desirability and feasibility of expansion of the state's aquaculture industry, and establish procedures and guidelines for the siting and conduct of marine aquaculture operations in the coastal zone.

The above are only some of the possible options for better incorporating marine aquaculture operations as a coastal-dependent use under the Coastal Zone Management Act. Amendments to legislation come about, however, only when those most responsible for and interested in a particular area of law and policy become convinced that some change is appropriate and necessary. To achieve this goal with regard to marine aquaculture and coastal planning, it will be imperative for the interest groups and agencies most concerned with aquaculture to establish good communication channels to federal and state coastal planning agencies, and to the interest groups that support and monitor the activities of coastal agencies, especially environmental organizations.

Management of Offshore Activities

Currently no federal framework is in place to manage the leasing of offshore submerged lands and waters for marine aquaculture purposes. The need for such a framework will become very apparent in the future when advances in technology allow marine aquaculture operations to go further offshore. At such a time, marine aquaculture operations may come into conflict with other users of the marine environment such as commercial fishermen, recreational fishermen, oil operators, marine transportation, military operations, and scientific research. The conflicts will be pronounced because aquaculture will represent an exclusive use of the water column—and, in some cases, of the submerged lands—in ocean areas that traditionally have been thought of as part of the public domain.

This scenario is not far in the future; as a matter of fact, these kinds of conflicts are already taking place in parts of the U.S. coastal ocean. Off Massachusetts, for example, environmental groups have initiated a lawsuit to block the development of a 47-square-mile offshore salmon farm located 37 miles off Cape Ann by American Norwegian Fish Farm (National Fisherman, 1991).

A framework is needed to provide an orderly process for the leasing and conduct of marine aquaculture operations to reduce the uncertainty that industry now faces in planning future activities. A management framework should have an environmental impact assessment requirement whereby potential environmental impacts can be identified and addressed; it should be aimed at identifying potential impacts on other users and evaluating appropriate strategies; it should provide a fair return to the public from the use of public waters, in the form of lease payments, royalties, and rents.

Because an orderly leasing framework for marine aquaculture in U.S. waters may attract predominantly foreign rather than domestic investment, the question arises of how to ensure maximum benefits to the U.S. marine aquaculture industry. In the case of commercial fisheries, priority was given to U.S. fishing vessels through the Magnuson Fishery Conservation and Management Act of 1976. Foreign fishing was allowed, in effect, only for the "surplus" that the U.S. industry could not harvest. In the case of the marine aquaculture industry, this goal could be accomplished through a variety of means, for example, by requiring that marine aquaculture firms operating offshore be at least 51 percent domestically owned.

SUMMARY OF POLICY ISSUES AND OPTIONS

Lack of National Leadership/ Insufficient Promotion Efforts

The federal government's policies to promote marine aquaculture have been confined to relatively easy and low-cost items (e.g., general policy statements of support, conduct of studies). Few tangible incentives have been provided to develop the industry, in contrast to the wide range of incentives and supports that agricultural businesses traditionally have received. In addition, no clear statement of the national interest in marine aquaculture has been articulated, particularly with reference to issues of international competitiveness and balance of payments.

Moreover, the level of R&D support for marine aquaculture has been low compared to agriculture and fisheries. This effort is carried out by the main federal agencies involved in marine aquaculture in a generally uncoordinated manner, notwithstanding the coordination mandate and mechanism called for in the 1980 and 1985 aquaculture acts. The JSA, as currently configured, has limited authority for effective interagency coordination other than on a voluntary basis.

A more proactive and forceful federal role will be needed for marine aquaculture to succeed in the United States. Changes are necessary in three major areas: (1) more extensive incentives and supports to industry; (2) an enhanced R&D effort; and (3) further centralization of authority and resources in the lead coordinating agency, the USDA.

Lack of a Solid Place in Coastal and Ocean Management Framework

Given the largely public nature of coastal waters and the public interest and management framework related to coastal lands (both public and private), a number of difficult issues arise regarding the privatization of public resources and spaces. On the one hand, the aquaculture entrepreneur needs to establish ownership over the product both to realize profits and to satisfy the institutions that provide essential financing. On the other hand, there are important public interest implications of leasing public waters for exclusive private use. Public waters traditionally have been common and open to all to use; privatization of any part of this public realm means that other direct uses (e.g., fishing, navigation) are affected, but it also raises concerns for the public as a whole. These public/private issues are complex and vary, to some extent, with the location and nature of the marine aquaculture enterprise.

Another problem is that no framework is in place for the management of marine aquaculture operations in federal waters (3–200 miles offshore). In the coastal zone, marine aquaculture is not included as a recognized use under the federal Coastal Zone Management Act.

The major opportunities for federal intervention to enhance marine aquaculture in the coastal ocean are thus threefold: (1) to create an orderly framework for the development of aquaculture operations in federal waters; (2) to influence the management of land and ocean resources under state regulatory control through available federal policy levers; and (3) to encourage the states, through a variety of technical assistance methods, to adopt and implement state aquaculture development plans.

Position of U.S.-Owned Operations Internationally and Foreign Ownership of Enterprises in the United States

The development of the fledgling U.S. marine aquaculture industry is inhibited directly and indirectly by several factors: (1) the presence of subsidized foreign competition through such programs as guaranteed loans, grants, and subsidies; (2) the presence of barriers to trade by other nations for exported U.S. fishery products; (3) the fact that aquaculture products imported into the United States can be grown and produced using practices and chemicals that would not be legal in the United States; and (4) the dumping of foreign products at "less than fair value." All of these factors may artificially depress prices to U.S. growers, causing significant economic losses.

Policies to mitigate these problems should be considered. Policies could include the use of countervailing and antidumping tariffs, restriction of trade of aquaculture products that are not produced in accordance with U.S. standards, export assistance through market intelligence and aid to U.S. growers, and preservation of the principle that the use of common-property U.S. marine resources should be for the benefit of U.S. citizens. The latter could be accomplished, for example, through a requirement that marine aquaculture operations in the United States be at least 51 percent U.S. owned.

Diversity of State Regulations

The federal government can play a technical assistance role and provide incentives for the states to streamline their bureaucratic processes with regard to marine aquaculture, and to develop and implement marine aquaculture promotion plans. Specifically, the federal government can undertake the following, through the JSA:

- assess the successes and failures of state and international experiences in the promotion and regulation of marine aquaculture;
- promote the development and implementation of statewide plans for marine aquaculture by drafting model regulation and guidelines;
- encourage the inclusion of marine aquaculture in the states' coastal zone planning processes;
- promote joint (local, state, and federal) intergovernmental review of marine aquaculture projects to ease the permitting burden on industry;
- promote naming of a lead state agency; and
- bring states together to share common problems and approaches, and to exchange technical information.

Fisheries Enhancement

Fisheries enhancement by marine aquaculture can be important for endangered, threatened, and overfished species. The private sector can contribute to this effort, providing cost-effective technical expertise and production capacity. The efficacy of expansion of enhancement activities to all endangered or otherwise threatened species needs to be examined. The nation is in need of a comprehensive policy to guide the expansion of fisheries enhancement activities that takes into account the advantages of allowing the private sector to play a major role. Existing policies often prohibit direct commercial competition or competition with public hatcheries, a case in which the competitor is also the regulator.

CONCLUSION

The key to finding the combination of measures that will improve the ability of marine aquaculture to function on an equal footing with other activities in the coastal zone is for the state and federal agencies involved in promoting and/or regulating these activities to work cooperatively. The USDA, which was designated as lead agency for aquaculture through the National Aquaculture Improvement Act of 1985 and which serves as chair of the Joint Subcommittee on Aquaculture, is the obvious and appropriate focus of leadership for such cooperative actions. However, it may take a more dynamic exercise of congressional oversight, through a congressional committee or subcommittee, to ensure that this mandate is implemented.

REFERENCES

Aspen Research and Information Center. 1981a. Aquaculture in the United States: Regulatory Constraints. Aspen Systems Corporation, Rockville, Md.

Aspen Research and Information Center. 1981b. A Directory of Federal Regulations Affecting Development and Operation of Commercial Aquaculture. Aspen Systems Corporation, Rockville, Md.

California. 1988. A Guide to California State Permits, Licenses, Laws and Regulations Affecting California's Aquaculture Industry. Interagency Committee for Aquaculture Development.

California Aquaculture Association. 1990. Strategic Plan.

Chasan, D.J. 1990. License to Pollute. The Weekly, Seattle, Wash. (November 8).

Cicin-Sain, B. 1982. Managing the ocean commons: U.S. marine programs in the seventies and eighties. Marine Technology Society Journal 16(4):6–18.

Cicin-Sain, B., and R.W. Knecht. 1985. The problem of governance of U.S. ocean resources and the new Exclusive Economic Zone. Ocean Development and International Law 15:289.

Conservation Law Foundation. 1990. Letter to Marine Board, October 16.

Corbin, J.S., and L.G.L. Young. 1988. Hawaii Aquaculture Resource Planning and Development: Past and Future. Proceedings, British Columbia Conference.

Davies, D.S. 1990. Allocating common property marine resources for coastal aquaculture: A comparative analysis. Ph.D. thesis, Science Research Center, State University of New York, Stony Brook.

DeVoe, M.R. 1990. Presentation to the committee. Hilton Head, S.C., August 6-8.

DeVoe, M.R., and A.S. Mount. 1989. An analysis of ten state aquaculture leasing systems: Issues and strategies. Journal of Shellfish Research 8(1):233–239.

Helm, L. 1989. Trouble down on the fish farm. *Seattle Post-Intelligencer*. (Dec. 18), p. B1, B4.

Hetrick, J. 1991. Alaskan aquaculture. *Water Farming Journal* 66(4):10-13 (April).

Institute of Medicine (IOM). 1991. Seafood Safety. Washington, D.C.: National Academy Press.

Joint Subcommittee on Aquaculture (JSA). 1990. National Aquaculture Forum Output, November 1987 (draft of September 7, 1990).

Joint Subcommittee on Aquaculture (JSA). 1991. Meeting minutes for September 12, 1990, December 18, 1991, and April 12, 1991, provided by R.O. Smitherman.

Knecht, R.W. 1986. In Ocean Resources and U.S. Intergovernmental Relations, M. Silva, ed. Boulder, Colo.: Westview Press.

Knecht, R.W., B. Cicin-Sain, and J.H. Archer. 1988. National ocean policy: A window of opportunity. Ocean Development and International Law 19:113-142.

Maine, State Planning Office. 1987. Establishing the Maine advantage: An economic development strategy for the State of Maine.

Maine, State Planning Office and Department of Marine Resources. 1989. An Aquaculture Production Strategy for the State of Maine.

Mayo Associates. 1988. The California hatchery evaluation study. Prepared for the California Department of Fish and Game. Sacramento.

McDonald, C.D., and H.E. Deese. 1988. Hawaii's ocean industries: Relative economic status. Proc. PACON 88, Pacific Cong. Mar. Sci. Technol. Honolulu. Pp. 16-20.

National Fisherman. 1991. Suit over offshore salmon farm plan. 72(2):8.

National Oceanic and Atmospheric Administration (NOAA). 1977. NOAA Aquaculture Plan. Washington, D.C.: U.S. Government Printing Office.

National Research Council (NRC). 1978. Aquaculture in the United States: Constraints and Opportunities. Washington, D.C.: National Academy Press.

Tiddens, A.A. 1990. Aquaculture in America: The Role of Science, Government, and the Entrepreneur. Boulder, Colo.: Westview Press.

U.S. Department of Agriculture/U.S. Department of the Interior (USDA/USDI). 1990. Final report of the USDA-USDI protective statutes workgroup. December (unpublished report).

Wenk, E., Jr. 1972. The Politics of the Ocean. Seattle: University of Washington Press.

4

Environmental Issues

Aquaculture, like traditional agriculture, creates environmental impacts. These impacts have received extensive scrutiny because marine aquaculture in the United States is relatively new and often conducted in public waters that are used and observed by many. Currently, four federal and numerous state and local agencies are involved in the regulation or monitoring of various aspects of aquaculture operations, including environmental impacts. The issues associated with environmental aspects of marine culture operations can be grouped into two broad categories:

1. impacts on the natural environment by the production systems, and
2. environmental requirements of the production systems, including impacts from and on other industries and interests (e.g., commercial fishing, recreation, human health).

ENVIRONMENTAL IMPACTS OF MARINE AQUACULTURE

Introduction

Concerns about the environmental impacts of marine aquaculture include such diverse issues as waste from cages or ponds, introduction of nonindigenous species or disease, the presence of infrastructure associated with culture operations in public waters, and genetic alterations of wild stocks through escapement of cultivated animals or intentional releases for stock enhancement.

Aquatic Plants

Of all the types of aquaculture operations, aquatic plant cultivation poses the least threat to the marine environment. Aquatic plant culture may be beneficial because it tends to counteract the potential detrimental effects of a variety of other coastal activities including terrestrial agriculture, sewage treatment, residential development, and fish or crustacean aquaculture. Aquatic plant culture using traditional rafting techniques relies on available dissolved nutrients and sunlight. Cultivated aquatic plants remove nutrients and limit eutrophication of the coastal environment. Aquatic plant culture employing rafting is insignificant in the United States, however, and rafting techniques may meet with resistance by boating interests or those concerned with the aesthetic aspects of a particular body of water.

Shellfish

The impacts of bivalve mollusk culture are also relatively innocuous, except in areas of highly intensive cultivation (e.g., mussel culture along the coast of Spain) (Figueras, 1989: Weston, 1991). Potentially adverse environmental impacts are similar to those for other species: (1) physical displacement or interference with other activities, (2) disturbances to natural phytoplankton communities (unlikely), (3) deleterious modifications of water quality through accumulation of wastes, (4) genetic contamination of wild stocks, and (5) introduction of species that compete with or are pathogenic to wild stocks (Weston, 1991). The majority of shellfish culture in the United States takes place in the public domain, particularly in estuarine and nearshore marine waters (Burrell, 1985; Lutz, 1985; Manzi, 1985). A small portion of this industry utilizes shore-based facilities.

Shore-based facilities typically house the hatchery and nursery components of businesses whose grow-out operations are in the estuary or nearshore coastal waters. The shore-based facilities rely to varying degrees on coastal water, which is pumped ashore. Effluents from shore-based facilities may be either enriched with cultivated microalgae produced for hatchery use, or partially depleted of naturally occurring phytoplankton and particulate matter that has been consumed in a nursery system. In either case, the effluent will have slightly elevated levels of metabolites, principally ammonia.

At present, virtually all shellfish production comes from open estuarine and nearshore waters, the use of which is generally regulated by the state. The degree of control exercised in shellfish cultivation varies dramatically. In many areas, shellfish cultivation is largely a matter of managing naturally recruited wild stocks. At the other end of the spectrum, more intensive operations deploy hatchery-reared spat into various types of floating or submerged hardware that provides predator protection and facilitates management and harvesting.

Such grow-out facilities may interfere with recreational or commercial activities (Burrell, 1985; Lutz, 1985; Manzi, 1985). Benthic communities may be impacted by submerged structures or nets, shell debris, or fecal sediment, food, and deposition from floating structures (Figueras, 1989; Weston, 1991). Their impacts on water quality and plankton communities are generally minor but may be measurable. Plankton is removed from the water and excreta (dissolved metabolites, feces, and pseudofeces) are then added to the water. Little, if any, change occurs in biochemical oxygen demand (BOD) and only a minor change in absolute dissolved oxygen concentration.

The source of the shellfish stock may be of concern if it is genetically different, represents a nonindigenous species, or is imported from areas that may harbor nonindigenous pathogens. The potential for an adverse effect from such stocks is increased by the fact that the cultivated crop is generally deployed directly into open waters, as opposed to pond or cage culture where there is some degree of confinement.

Shrimp

Virtually all shrimp farming in the United States employs ponds (Chamberlain, 1991; Hopkins, 1991; Pruder, 1991), although several ventures have cultured shrimp in environmentally controlled greenhouse-covered tanks (Salser et al., 1978). Some ponds are actually previously impounded wetlands (Whetstone et al., 1988), and few attempts have been made at culturing shrimp in net enclosures. Typically though, the ponds are constructed on high ground adjacent to a supply of seawater. Estuarine water is as satisfactory as ocean water. Saline groundwater may be satisfactory if the ionic composition is similar to that of seawater. A second component of the shrimp farming industry is hatchery production of postlarvae for stocking ponds. Hatcheries use relatively little water, but it must be of near-oceanic quality.

There are several areas of concern relative to environmental impacts of shrimp farming. These concerns can be broadly categorized as (1) genetic-related threats to indigenous species; (2) disease-related threats to indigenous species; and (3) threats related to water quality degradation in the effluent receiving stream. The probability of these impacts varies among the three geographic areas in which U.S. shrimp farms are concentrated: Texas, South Carolina, and Hawaii.

U.S. shrimp culturists rely almost exclusively on a nonindigenous species *Penaeus vannamei* (Rosenberry, 1990; Wyban and Sweeney, 1991). Postlarval seedling shrimp for stocking ponds are obtained from commercial hatcheries in the United States and Latin America. There is some concern that this nonindigenous species could become established and displace indigenous species, particularly the Atlantic and Gulf of Mexico white shrimp *(P. setiferus)*. The possibility of hybridization among these species has been raised, but it does not appear to be a realistic concern. In response to concerns about the importation of nonindigenous species, research has focused on the development of native species that may have marine aquaculture potential (Sandifer et al., in press). However, the process of domestication of shrimp stocks through selective breeding of indigenous species could impact the genetic diversity of wild stocks were there to be large-scale or continuous escapement of the domesticated animals. Thus, it is conceivable, although unlikely, that a highly selected line of an indigenous species could have as great or greater impact than imported nonnative species. This is the same concern expressed for hatchery stocks of salmonids.

Although shrimp diseases are poorly understood at present, some diseases appear to be associated with particular geographic areas, species, or aquaculture operations. The pathogen of most concern is the infectious hypodermal and hematopoietic necrosis (IHHN) virus, which has been shown to cause stunting, deformities, reduced growth rates, or mortality in several species (Browdy et al., 1990; Kalagayan et al., 1990). The response to IHHN infection is highly species-specific. Although no cases of aquaculture operations causing disease outbreaks in adjacent wild stocks have been documented, continued vigilance, escapement prevention, and shrimp disease research are essential if this industry is to continue to develop in the United States.

For shrimp culture in the United States to be competitive in the worldwide shrimp marketplace, farms must use intensive production technology (Sandifer, 1988; Wyban and Sweeney, 1991). The concentration of pollutants in the effluent increases with intensification due to higher feeding rates. The potential environmental effect of shrimp farm effluent is increased eutrophication of the receiving stream through nutrient addition if proper dilution rates are not mandated (Brune, 1990). Water quality parameters of concern include BOD, ammonia, and suspended solids.

Modeling of shrimp pond effluents based on the level of intensification and water exchange is now possible (Brune, 1990; Brune and Drapcho, 1991). Coupled with existing models of effluent dilution and ultimate oxygen decline in complex tidal receiving streams, this gives the farmer or regulator a powerful tool with which to predict environmental impacts. Delineation of an acceptable impact from an unacceptable adverse impact is still not clear, however. Research is currently under way to reduce the BOD and nutrient loads of effluents from intensive shrimp farms, and this is an obvious area in which technological advances could improve the possibilities for growth of shrimp farming in the United States (Sandifer et al., 1991a, b).

Finfish

Although the culture of mollusks, fish, and crustaceans accounts for most of the production by the U.S. marine aquaculture industry, environmental concerns in some parts of the country are focused on floating cages used for salmon culture. To date, few long-term studies have been conducted on this subject in the United States and Canada; however, much research and environmental monitoring of net pens has been done over the past 20 years in Europe and Japan. Many of the early aquaculture projects were located in semienclosed areas with poor water exchange; consequently, the first studies on environmental effects showed significant but localized impacts (Rosenthal, 1985). Recent comprehensive studies suggest that the environmental impacts of properly sited cages can be alleviated through the development of improved management and production systems (Gillespie, 1986; Weston and Gowan, 1988; Paramatrix, 1990; Cross, 1990). However, the use of coastal habitat by aquaculture facilities may impinge on native species' habitat and cause reductions in the populations of the native organisms.

Impacts From Waste

Wastes from culture operations can have a variety of environmental impacts. Two primary concerns relate to water quality and benthic ecology.

Water Quality

Finfish or shrimp in ponds or tanks dramatically affect water quality primarily through excretions from feed inputs. Water quality differences between inlet and effluent waters are a function of the loading of fish, the water exchange rate (retention time), and the feeding rate. When water retention time is long, feed inputs are digested, either by the fish/shrimp crop or by microbial digestion, and mineralized.

Major end products in the digestion process are dissolved nitrogen and phosphorus species (mainly ammonia and orthophosphate) and particulate matter. Only 20 to 30 percent of the nitrogen input as feed is assimilated into fish tissue (Krom et al., 1985; Porter et al., 1987). Ammonia is the primary end product excreted by fish, crustaceans, and mollusks (Campbell, 1973), and its release generally is proportional to the feeding rate (Colt and Armstrong, 1981).

These digested end products may be reassimilated by phytoplankton, protozoans, bacteria, and fungi. Such organisms have short life spans, and on their decay, nutrients are again mineralized into dissolved or particulate debris forms. This cycle continues until the material is finally (1) flushed from the system with water exchange, (2) deposited in more stable sediments, (3) volatilized to the atmosphere, or perhaps (4) assimilated by organisms large enough to be consumed by the fish/shrimp crop. When material that was once feed input exits the pond with water exchange, it is an effluent "pollutant." Water exchange is typically the greatest source of nitrogen loss from the system (Daniels and Boyd, 1989).

Sedimentation of solids and sludge formation may be an important sink for nitrogen and other pollutants. Sludge accumulations ranging from 11 to 38 percent of the feed applied have been reported, the differences being attributed to sludge digestion because of variable holding times (McLaughlin, 1981). It has been suggested that sludge accumulations decrease available habitat for shrimp, reduce the density of benthic food organisms, and cause direct toxicity due to hydrogen sulfide and other anaerobic metabolites (Chamberlain, 1986). However, these impacts have not been documented, and healthy shrimp can be found in sludge deposits. In addition, populations of benthic organisms are grazed nearly to extinction in intensive shrimp culture ponds (Hopkins et al., 1988a,b), and very little hydrogen sulfide has been found free in the water column (Ellis, 1990). The more important impact of sludge accumulation may be the sludge digestion processes that demand oxygen and release bound nitrogen back into the system. If sludge is discharged with the exchange water, it degrades the quality of effluent by elevating concentrations of BOD and solids.

The fish/shrimp crop is a major source of oxygen depletion in densely stocked tank systems, and reoxygenation is provided via aeration equipment. At the stocking densities typical of pond culture, the primary oxygen consumers are the decay and photosynthetic organisms in the water column and pond bottom. The higher the pond feed input, the higher must the supplemental aeration rate be to maintain adequate dissolved oxygen at night (Hopkins et al., in press). The effluent dissolved oxygen is generally as high as that of the receiving body in aerated tanks and ponds. Pond effluent dissolved oxygen may be higher than that of the receiving body during the day due to photosynthetic activity.

Cage systems are not artificially aerated and have rapid water exchange. The water passing through the pen typically has a slightly lower dissolved oxygen and slightly elevated ammonia concentration. The mass balance of feed input and pollutant output is equal, less the small amount assimilated into fish tissue. However, the dilution rate is extremely high in pen culture, and much of the secondary food decomposition occurs outside the pen, as in a rapidly flushed tank system. Model predictions and field measurements downstream from salmon cage farms in Puget Sound typically show a decrease of less than 0.3 milligram (mg) per liter in oxygen (Weston, 1986). Salmon require high oxygen levels; therefore the impact of lowered oxygen levels is self-limiting.

The principal nutrient contributed to the environment from cages is nitrogen. Salmon annually produce between 0.22 and 0.28 gram (g) of dissolved nitrogen (mostly ammonia) per kilogram of fish (Gowen and Bradbury, 1987). This nitrogen results in an increase of approximately 0.02 mg/liter of ammonia downstream from the average salmon farm (Weston, 1986), a small fraction of the Environmental Protection Agency (EPA) water quality standards for ammonia. Salmonids are extremely sensitive to ammonia, so this impact, like oxygen reduction, may be self-limiting with salmon. Recent comprehensive studies by Paramatrix (1990) in the United States and Gillespie (1986) in Canada on salmon net-pen farms conclude that water quality impacts are slight, localized, and reversible. Similar opinions were expressed in presentations to the committee (Gowen and Rosenthal, 1990).

The composition of waste from cultured fish differs little from that contributed naturally by wild fish, but it differs significantly from that of warm-blooded animals. The effect of culture operations on coliform, and particularly fecal coliform bacteria, is a water quality concern. A better understanding is necessary, including a clearer differentiation between fecal and total coliform (ICES, 1988a).

Plankton

Shellfish tend to remove phytoplankton from the water during filter feeding, which may decrease the food supply for other animals. Counterbalancing this is the fact that marine plankton growth is often nitrogen limited. As a result, fish farms have the potential to cause or exacerbate plankton blooms by virtue of the nitrogen produced. The recent increase in awareness of toxic plankton blooms worldwide has raised concerns that aquaculture might contribute to the problem (Whiteley and Johnstone, 1990). Correlations between aquaculture and harmful blooms have been documented in Japan where intensive culture of finfish and shellfish occurs in poorly flushed bays (Nose, 1985). Other than in Japan, few, if any, cases have been documented in which aquaculture has caused algal blooms (Gowen and McLusky, 1990). Marine aquaculture can be the victim of plankton blooms (Saunders, 1988; Shumway, 1990). Toxic blooms sometimes cause closing of shellfish beds and, in some cases, can be lethal to fish.

Benthos

Accumulation of wastes can alter benthic ecology and modify the chemistry of growing waters. Net-pen marine aquaculture operations typically result in large amounts of solid wastes, including feces and uneaten food from fish pens, and pseudofeces and shell debris from mollusk culture. A portion of the solid waste produced in tanks and ponds is digested in situ when water retention times are long. Thus, the effects of their effluents on benthos may be less than those from pen culture systems. Settleable waste from culture operations may alter the ecosystem by changing the physical and chemical environment or by changing or reducing the numbers and species resident beneath net pens or downstream from effluents. Solid waste is estimated at between 0.5 and 0.7 g for each kilogram of fish produced (Paramatrix, 1990). Although the quantity of waste is significant, it tends to accumulate beneath the pens only in sites of less than 15 meter (m) depth and low current velocities (Weston, 1986). Studies on existing net pen operations in North America show that even on large farms where accumulations do occur, the impact is confined to an area roughly 30 m around the pens (Weston, 1986; Cross, 1990; Paramatrix, 1990).

Models based on current velocity, depth, loading rate, and other factors are now available to select sites where impacts of new farms will be minimal (Weston and Gowen, 1988). The same observation holds true for pond and tank systems. If current velocities at the effluent discharge site are high, dispersal and dilution minimize any effects on benthos. In addition, evidence indicates that benthic impacts are rapidly reversed when net pens are removed (Dixon, 1986).

Mollusk culture also can result in accumulation of waste (ICES, 1988a). Shell rubble directly below intensive mussel and oyster culture systems can result in significant effects on the benthos directly beneath such operations if the rubble is not collected, the culture site is not selected to minimize the impacts, or the site is not mobile.

Accumulation of anoxic sediments has occurred in some shallow bays in Japan as a result of mussel and oyster culture (Nose, 1985). Accumulation of anoxic sediments has also been noted in pond culture of oysters where phytoplankton densities are high and large amounts of pseudofeces are being produced. Although shell rubble does alter the benthos, it does not increase BOD, tends to stabilize sediments, and may provide settlement or attachment sites for wild shellfish.

Regulation of Discharges

Aquaculture facilities can produce sizable quantities of waste and discharge large volumes of effluents to surface waters. Therefore, aquaculture operations (along with agricultural operations) are faced with growing environmental regulatory scrutiny. Although much of the regulatory activity has come from state and local sources, a number of federal statutes and regulations directly impact the management of aquaculture wastes and effluents.

The Clean Water Act

The Clean Water Act (CWA) of 1977 (40 CFR) focuses on the protection, restoration, and maintenance of the chemical, physical, and biological integrity of the nation's waters. The CWA authorizes the issuance of federal National Pollution Discharge Elimination Systems (NPDES) permits for point source discharges (including delegation of the federal permit program to the states), and the development of areawide waste treatment management plans, including best management practices (BMPs) for nonpoint sources of water pollution. Under the general NPDES permit regulations (40 CFR Part 122), "concentrated aquatic animal production facilities" are considered *point sources* requiring NPDES permits for discharges into waters of the United States. "Concentrated aquatic animal production facilities" are defined as a hatchery, fish farm, or other facility that meets the criteria in appendix C of the Clean Water Act, or any such facility that the director determines is a significant contributor of pollution to waters. The criteria provided in appendix C generally include commercial-size marine aquaculture fish farms or other facilities that "contain, grow, or hold cold water aquatic animals in ponds, raceways, or other similar structures which discharge at least 30 days per year."

Therefore, aquaculture production facilities that meet these criteria or are found to be significant contributors to water pollution are subject to NPDES permits under the Clean Water Act. Moreover, states may place additional requirements on these discharges. Because many states have been delegated the authority to issue federal NPDES discharge permits, some states issue joint federal NPDES/state permits.

Some aquaculturists have suggested that aquaculture effluents should be treated as *nonpoint sources of pollution* (a different category under the CWA, analogous to runoff from agricultural fields as contrasted with discharges from a feedlot), which are presently less stringently regulated under the CWA. However, states and federal agencies are currently in the process of imposing stricter regulations on all nonpoint sources of pollution. For example, a study convened by the EPA administrator recently recommended that "the states and the federal government augment voluntary programs with increased use of regulatory authority for reduction of nutrient loadings of the Chesapeake Bay [from agricultural runoff]" (Chesapeake Bay Program, 1991).

Managing Wastes and Effluents

Aquaculture wastes and effluents can be managed through well-designed and operated recycling programs that beneficially utilize the "waste" products as resources. Such programs include utilizing the organic solids to improve or fertilize soil, as animal feed supplements, or using the wastewater as irrigation water, cooling water, or for recycling to the same or other aquaculture production systems (Mudrak, 1981). Well-managed beneficial use practices can help conserve water supplies and significantly reduce the volume requiring disposal (Rosenthal, 1985).

Impacts From Introduction of Nonindigenous Species

Agricultural production in the United States, as in most other countries, relies almost entirely on the cultivation of introduced species. Today, most animal and plant foods come from a relatively few species that are grown where suitable environments exist. Aquaculture also relies on introduced species that have excellent market value and acceptance and that are amenable to cultivation.

Introductions of nonindigenous species raise the possibility that the introduced species will (1) compete with native organisms for existing ecological niches, (2) alter the food web, (3) modify the environment, (4) introduce new diseases, and/or (5) dilute native gene pools through interbreeding, hybridization, or especially, ecological interaction. The biggest problem associated with nonnative introductions is lack of information about the short- and long-term impacts of the introduced species on its new environment. Unanswered questions about the long-term effects of introduced species include the following (Seter, 1990):

- *Competition via interference or exploitation*: Will the introduced species occupy a previously untapped niche or compete with native organisms for existing niches?
- *Predation*: What impact will the introduced organism have on the surrounding ecosystem? Will food webs be permanently altered?
- *Environmental modifications*: Will water quality be affected? Will the introduced species physically alter its surroundings?
- *Hybridization*: Will inhibition of reproduction or, at the other end of the spectrum, interbreeding dilute or degrade native gene pools, reducing the potential for future benefits from wild gene stocks?

The transfer of aquatic species can occur through unintentional as well as intentional acts. Means of transfer are varied (Chew, 1990) and include the following:

- transfer by water traffic on or in ships, especially ballast water (e.g., the recent introduction of the zebra mussel to the Great Lakes (Griffiths et al. 1991));
- escape or release of organisms transferred for other purposes, such as confined culture, direct consumption for food, or use as ornamentals (live

crabs, lobsters, and mollusks, are routinely transported worldwide, as are fish and invertebrates for the aquarium industry);

- accidental transfer of a secondary species associated with the transfer of a target species (i.e., organisms transferred in or on their hosts); and
- deliberate transfers and introductions for culture or fisheries enhancement.

Examples of nonaquaculture sources of introductions include the transport of nonindigenous species on ship hulls and in ballast water, which led to the introduction of the Australian barnacle and Chinese mitten crab in Europe (Rosenthal, 1980); the introduction of Pacific species into the Atlantic and vice versa through the Suez Canal (Vermeij, 1991); the inadvertent or purposeful release of a great variety of aquarium fish and plants, and the shipment and frequent release of live bait organisms (Courtenay, unpublished manuscript).

Genetic Impacts

Genetic changes of wild stocks can result from (1) straying of anadromous fish released for fisheries enhancement or ocean ranching, (2) escape from confinement facilities, or (3) purposeful release of cultured fish (Sattaur, 1989). Some investigators suggest that the potential loss of genetic diversity in a species can negatively affect its present condition and, more important, potentially affect the species' ability to adapt to a changing environment (Hindar et al.). Other workers in the field, however, consider the genetic effects of large-scale releases with a more benign—and even positive—attitude. For example, Mathisen and Gudjonsson (1978) argue against a purist opposition to mixing gene pools of Atlantic salmon for release.

Genetic issues apply to all cultured species, but a recent controversy involves private salmon ocean ranching and public fisheries enhancement along the West Coast, where hundreds of millions of hatchery fish are released yearly (Waples et al., 1990). Some hatchery fish, both private and public, will stray as will some fish from natural runs. The rate of straying of hatchery fish is influenced by release strategy and possibly by hatchery strategy; however, the influence of stray rate on the actual genetic impact has not been adequately evaluated. Interpretation is confounded because the frequency of the gene flow associated with natural straying is unknown.

At present, the greatest environmental concern appears to be the potential for overwhelming the wild gene pool with the more restricted gene pool of a hatchery stock through repeated and massive stock releases, as with salmon in the Northwest (Hetrick, 1991; Hindar et al., 1991). Clearly, the gene pool of any stock that is reared in a hatchery and originates from the spawning of fish that return to the hatchery will be altered over time. Concern also exists that escapement of stock from confined culture operations will lead to weakening of the wild stock. For this to occur, large numbers of animals must escape, survive in an unfamiliar environment, compete, and breed successfully with wild stocks. The scenario may be now in certain parts of Norway where salmon cage (net-pen) farming is intensive and only remnant wild populations remain (NASCO, 1990). The major risk from these hatchery programs is ecological interaction of hatchery and wild fish, resulting from overstocking natural waters or allowing wild stocks to become severely depleted (Sattaur, 1989).

Interbreeding of fish that escape from net pens with truly native wild salmonids in the continental United States is unlikely (Gillespie, 1986; Paramatrix, 1990). On the East Coast, the native Atlantic salmon all but disappeared in the 1940s and was replaced with a variety of Canadian strains by the U.S. Fish and Wildlife Service. Public hatcheries continue to stock progeny of these strains because natural reproduction is very low. However, hatchery fish are different from wild fish, and the practice of stocking hatchery fish to augment populations where native stocks are in decline is under intense scrutiny. On the West Coast, estimates of cage escapees are insignificant in number compared to the hundreds of millions of fish released by public mitigation and stock enhancement programs. The possibility does exist, however, for any cage escapees to interbreed with the introduced Canadian strains. As stock/strain identification procedures are refined, the problem can be better evaluated.

During the past 15 years, interest has grown in the enhancement of wild stocks through the release of hatchery-reared fish and shellfish. Millions of young red drum, striped bass, sturgeon, and oysters produced in hatcheries have been stocked in natural waters, and the release of other species such as tarpon, snook, and red snapper is under consideration. The fecundity of these species could potentially lead to the release of an overwhelming number of progeny with limited parentage, which might result in reduced genetic diversity of the population.

Risks are involved with the introduction of new strains through escapement or planned release of hatchery-reared fish. The degree of risk has not been determined. Research is needed to provide a thorough understanding of the risk and of how to manage enhancement programs most effectively. Care must be taken to preserve native stocks and avoid unplanned reduction of genetic diversity (Nehlsen et al., 1991).

Disease Transfer

Another widespread concern is that disease from farmed species might be transferred to wild fish or shellfish or that new diseases will be introduced through imported eggs, larvae, or juveniles. Farmed fish or shellfish could also serve as a reservoir for disease organisms (Munro and Wadell, 1984). The major emphasis with regard to the possible transfer of disease is on preventing the spread of untreatable diseases (viral or myxosporidal); treatable diseases (bacterial, fungal) are of less concern. Strict regulations involving the quarantine and testing of species for diseases and parasites prior to their introduction, are important, as the discussion below points out. Many disease-related problems with aquaculture appear in conditions of confinement. These diseases often do not manifest themselves in the natural environment where stress factors are reduced.

Most cases of disease transfer from cultured to wild stocks occur in conjunction with introductions of nonindigenous species or populations. Weston's (1991) review of the literature indicates that at least 48 species that are parasitic on freshwater fish have been transferred among continents via the importation of live or frozen fish and that the IHN (infectious hematopoietic necrosis) virus of trout has been spread throughout the northwestern and north central United States and into Japan via shipments of infected organisms. Other examples include the apparent introduction of "crayfish plague" to Britain by farmed crayfish originally imported from North America (Thompson, 1990) and the transmission of predators and parasites of bivalve mollusks via shipments of Pacific oyster seed and other bivalves (Rosenfield and Kern, 1979; Chew, 1990). Introduction of nonnative specimens of native species may also be accompanied by predators, parasites, and diseases (e.g., the introduction of a sacculinid barnacle parasite of mud crabs to the Chesapeake Bay via shipments of American oysters from the Gulf of Mexico (Van Engel et al., 1966).

Many states have implemented some form of disease testing and certification programs for animals being imported across state lines. Such programs often test mainly for diseases already present in the area, and established programs are limited almost entirely to freshwater species. Salmon egg and smolt importation is highly regulated by the Fish and Wildlife Service and state agencies. In some states, a quarantine period is required for salmonids prior to introduction. Unfortunately, many of the state inspection and certification programs for saltwater species have insufficient capability to conduct comprehensive inspections.

Major technological and institutional problems remain regarding diagnosis and control of diseases of marine fish and shellfish. A number of states and institutions have fairly broad expertise in the diagnosis of diseases of established cultured fish species such as trout, salmon, and catfish, and "disease-free" certification programs generally are well established for them. To a considerably lesser extent, certification protocols also exist for oysters, clams, and shrimp. The knowledge base for diseases of marine crustaceans (e.g., shrimp), however, is relatively poor. Diagnostic procedures are quite limited; federal certification procedures/laboratories are nonexistent; and qualified state certification operations exist in, at most, a few states.

Fish and shellfish from U.S. capture fisheries must meet only public health criteria, even if they are being harvested for holding or shipment live to other areas (e.g., oysters, clams, scallops, mussels, lobsters, crabs). Routine shipments of live shellfish and crustaceans intended for direct sale to consumers or for use as bait are seldom, if ever, examined for diseases, parasites, or accompanying organisms. Nevertheless, such shipments may be significant potential sources of disease (IOM, 1991). Nor are frozen and fresh seafood products imported into the United States generally inspected for disease, although they may serve as an avenue of disease transfer to native stocks.

Regulation of Fish Movement

The federal government regulates movement of nonindigenous species through the Lacey Act (P.L. 97–79, as amended in 1981), and the states exercise varying degrees of control over the use and introductions of exotic nonindigenous species. Requirements include importation permits, an environmental risk report, inspection certifying the lack of disease, and in some cases, a disease history of the stock.

In the majority of states, introduction of nonnative species requires authorization from a state conservation agency (King and Schrock, 1985). In many cases, standards for private hatcheries and farms exceed those applied to public hatcheries (Hicks, 1989). Importation of salmonid eggs and fish is highly regulated by federal and state agencies. Importation of fish is prohibited under most circumstances, and egg importations are restricted to inspected stocks from specific regions. Salmon egg and smolt importations are highly regulated, and in some states a quarantine period is required prior to introduction.

The International Council for Exploration of the Sea (ICES), of which the United States is a member nation, has compiled a detailed and comprehensive protocol for introduction of exotics (ICES, 1984), which has been suggested as a guide for all planned introductions of marine species (Sindermann, 1988). In the context of disease control, this protocol requires careful screening for disease organisms and holding brood stock in quarantine until the production of first-generation organisms. This protocol was used successfully in an introduction of eastern bay scallops from the United States to Canada. However, a problem limiting practical implementation of the disease protocol is that insufficient knowledge is available about the diseases or parasites of importance or about the diagnostic tools for most species (Sindermann, 1988). Lightner (1990), referring to the ICES protocol and the FAO (1977) guidelines, stated that "for these guidelines to work, adequate quarantine facilities and qualified diagnosticians must be available."

The problem is illustrated by the example of a penaeid shrimp disease in Hawaii. A strict quarantine system was established for the introduction of nonnative shrimp species based on the ICES protocol. The protocol was targeted especially to prevent introduction of the IHHN and other viral pathogens. Nevertheless, despite strict controls and apparently excellent compliance by the aquaculture industry, the IHHN virus was diagnosed in a Hawaiian population of *Penaeus stylirostris* in 1987 and in *Penaeus vannamei* in 1989 (J. Brock, Hawaii Department of Land and Natural Resources, Aquaculture Development Program, personal communication, 1990). The disease is also found virtually everywhere these species are cultured.

Impacts of Feed Additives

Antibiotics may be added to fish feed to reduce mortality from bacterial fish diseases such as vibriosis and furunculosis. These antibiotics are used in marine aquaculture as prophylaxis and as therapy for disease outbreaks. In other animal production operations, such as for cattle and pigs, antibiotics are frequently used on a continual basis to prevent disease and enhance growth (NAS, 1980). At present, only three antibiotics are approved for use during disease outbreaks on fish farms in the United States—oxytetracycline (OTC), sulfamerazine, and Romet 30, a sulfa drug. Of these three, OTC is by far the most commonly used antibiotic.

Concerns about antibiotics stem from three potential environmental effects (Whitely and Johnstone, 1990):

1. development of drug-resistant strains of bacteria,
2. accumulation of antibiotics in sediments and subsequent inhibition of microbial decomposition, and
3. accumulation of antibiotics in fish and shellfish.

The first two concerns are based on actual occurrences under specific conditions. Aoki and Kitao (1985) found drug-resistant bacteria in the effluent of an intensive culture fish pond in Japan. Jacobsen (1989) reported OTC in the sediments beneath net pens in Norway, and drug resistance was transferred from a fish pathogen to a human pathogen in vitro and at temperatures as high as 36°C (Toranzo et al., 1984).

The frequency of drug-resistant bacteria does increase as a result of antibiotic use in animal feed, and this resistance can be transferred to human and animal pathogens (Wright, 1990). However, the evidence remains circumstantial that human health is threatened even under the continual use of antibiotics in livestock operations over many years (Walton, 1988). Marine fish culturists in the United States use antibiotics only on a limited basis. For example, net-pen growers may use OTC for two or three treatments of 10 days each during the year. Antibiotics are used only when necessary.

Accumulation of an antibiotic in sediment depends on many factors, including its solubility, half-life, and concentration in seawater. OTC is highly soluble in seawater and has a short half-life (Jonas et al., 1984). Austin (1985) calculated that under a worst-case scenario, the highest antibiotic levels in receiving waters would correspond to a dilution of 1:50,000,000. They concluded from this finding that the release of pharmaceutical compounds from fish farms was unlikely to pose an environmental problem.

Several studies have demonstrated that shellfish did not accumulate antibiotics in their tissue above the concentration in the surrounding water (NAS, 1980; Tibbs et al., 1988).

A fourth concern about antibiotics is the possible impact on human consumers from antibiotic residues in fish. The risk is greater for imported fish because the kinds of antibiotic treatments and their duration on U.S. fish farms are regulated more stringently in the United States than elsewhere. The time during which antibiotic residues remain in trout muscle depends largely on water temperature. For salmonids given OTC, recommended withdrawal times are 60 days at a water temperature of 12°C and 90 days at 6°C (Jacobsen, 1989). At present, no inspection procedures are in place for imported fish, but cooking destroys most OTC residues in salmonids (Herman et al., 1969). Little information is available on clearance times and residues in nonsalmonid farmed fish. More understanding is needed of the potential deleterious effects on the environment from treatment of disease in the culture operation (e.g., pesticide treatment for fish lice infestment in net pens).

The Food and Drug Administration (FDA) has recently adopted a stringent policy on the use of unapproved drugs in aquaculture, a policy that could have profound impact on standard aquaculture practices. The policy requires producers and researchers to obtain approval from FDA for investigational use before they can use any drug not formally approved. The process for obtaining formal approval of a drug is likely to involve a time-consuming and expensive process. The FDA points out that current federal and state funding for drug development research is inadequate to meet the needs of the aquaculture industry, and suggests that congressional appropriations be allocated for this endeavor (Water Farming Journal, 1991).

ENVIRONMENTAL REQUIREMENTS OF MARINE AQUACULTURE

Marine aquaculture has as a basic environmental requirement, accessible water of suitable temperature, quality, and quantity. Varying amounts of water exchange are necessary, depending on the species. Also of importance is the selection of a site where stock can be protected from weather extremes and from human or animal interference. Marine aquaculture is highly vulnerable to external pollution by domestic and industrial wastes, oil and chemical spills, and other discharges that may originate from sources remote from the culture operation but be carried to it by tides and currents.

The discharge of toxic industrial waste is a hazard to marine aquaculture because shellfish and seaweed are particularly vulnerable to heavy-metal pollution as well as to pollution from synthetic organic compounds. The cultured organism can concentrate mercury, lead, cadmium, arsenic, polychlorinated biphenyls (PCBs), and other toxic compounds to such an extent that it is altered, killed, or rendered unsafe for human consumption.

By far, the greatest impact on aquaculture from pollution, however, has been the closure of both natural and cultivated shellfish beds due to pollution from animal and human wastes. The nutrients in domestic wastewater, whether it is treated or untreated, also may induce blooms of toxic or otherwise harmful algae, for example, by increasing the concentrations of primary nutrients (inorganic nitrogen, phosphorus), and through organic overloading.

Mollusks

Shellfish aquaculture requires approved (waste-free) marine or brackish water with suitable food organisms, specific depths and temperatures, and low turbidity. Sites are limited because shellfish are vulnerable to external pollution by industrial, municipal, and agricultural wastes owing to their feeding habits. Major closures of both natural and cultivated shellfish beds have been caused by the presence of bacteria from domestic sewage. This problem has resulted in the elimination of one-half or more potential culture sites in many regions, including the Chesapeake Bay and San Francisco Bay. Closures are also caused by nonpoint sources of pollution. For example, many locations have enforced automatic closures after rainfalls of preset intensity and duration.

In addition, shellfish may also become contaminated with poisons by ingesting toxic microorganisms from the water, which makes them unsafe for human consumption due to the danger of paralytic shellfish poisoning (PSP). In California, a mussel watch program that includes participation by aquaculturalists monitors for toxic conditions to regulate closure of public gathering grounds as well as to suspend harvest at culture facilities.

Another concern is the possible transfer of human pathogens from polluted growing water to the shellfish and then from the shellfish to humans who eat them raw. Pathogens of concern are polio, hepatitis A, and Norwalk viruses, as well as *Vibrio* spp. and other enteric bacterial pathogens (Richards, 1988). Such pathogens generally originate in domestic wastewater. The current standard used for monitoring shellfish and culture waters for the purpose of public health protection is recognized as inaccurate and inadequate. The fecal coliform test does not measure the relevant microorganisms (viral and bacterial pathogens) and does not provide a useful index of sewage pollution. Fecal coliforms have been found to reproduce in the aquatic environment and are produced and released by aquatic birds, domestic animals, and wildlife, as well as by humans (IOM, 1991).

Finfish and Shrimp

Marine finfish farms in the United States are located nearshore (cages) or onshore (tanks, raceways, and ponds). Species requirements sharply limit the number of suitable sites. For example, a site for salmon cages must have unpolluted water at least 10 m deep, a water temperature of 0–18°C, current between 10 and 100 centimeters per second, and protection from severe weather. A site for culture of red drum or shrimp requires a location where seawater can be effectively pumped to the facility. Prices of suitable land are generally determined by residential or commercial interests, which limit the economic feasibility of an aquaculture operation. Regulatory constraints on aquaculture effluents also present major problems in site selection. For example, many miles of coastline in Hawaii are zoned to prohibit discharges of any kind (Ziemann et al., 1990).

For anadromous fish, large amounts of fresh water are usually required in early life stages. Hatchery sites for anadromous finfish on the West Coast are limited, and there are restrictions on groundwater use in the lower Mississippi delta and the Atlantic coastal plain. Seawater intrusion into freshwater aquifers is becoming more prevalent, resulting in increased restrictions on the use of water from these aquifers.

RESOLVING ENVIRONMENTAL PROBLEMS

In some cases, the mitigation of environmental problems associated with marine aquaculture may be possible through improved understanding of biological and ecological factors involved in culturing various marine species, and through engineering and technology solutions that allow new approaches to siting and to culture operations. These options are explored in detail in Chapter 5. Some of the state and federal policy issues discussed in Chapter 3 are also relevant to environmental issues, and changes in management and regulatory approaches may alleviate environmental controversies.

The aspects of environmental issues that involve public attitudes and values may be addressed through active efforts at educating both the public and policymakers about the benefits of aquaculture and the prospects for alleviating some of the most serious environmental impacts. Solutions to the environmental problems constraining marine aquaculture will involve approaches that combine technological "fixes" with improved regulatory and management structures, as well as public education about the value of marine aquaculture to the nation.

REFERENCES

Aoki, T., and T. Kitao. 1985. Detection of transferable R plasmids in strains of the fish-pathogenic bacterium *Pastuerella piscicida*. Journal of Fish Diseases 8:345-350.

Austin, B. 1985. Antibiotic pollution from fish farms; Effects on aquatic microflora. Microbiological Sciences 2(4):113-117.

Browdy, C.L., J.R. Richardson III., C.O. King, A.D. Stokes, J.S. Hopkins, and P.A. Sandifer. 1990. IHHN virus and intensive culture of *Penaeus vannamei*: Effects of stocking and water exchange rates on production and harvest size distribution. World Aquaculture Society, World Aquaculture 90, Abstract T17.3.

Brune, D.E. 1990. Reducing the environmental impact of shrimp pond discharge. American Society of Agricultural Engineers, ASAE Paper No. 90-7036. St. Joseph, Mich.

Brune, D.E., and C.M. Drapcho. 1991. Fed pond aquaculture. Pp. 15-33 in Aquaculture Systems Engineering: Proceedings of the World Aquaculture Society and American Society of Agricultural Engineers Jointly Sponsored Session. American Society of Agricultural Engineers, St. Joseph, Mich.

Burrell, V.G., Jr. 1985. Oyster culture. Pp. 235-273 in Crustacean and Mollusk Aquaculture in the United States, J.V. Huner and E.E. Brown, eds., Westport, Conn: AVI Publishing Co.

Campbell, J.W. 1973. Nitrogen excretion. In Comparative Animal Physiology, C.L. Prosser, ed. Philadelphia, Pa.: W.B. Saunders.

Chamberlain, G.W. 1986. 1985 Growout research. Coastal Aquaculture 3(2):7-8.

Chamberlain, G.W. 1991. Status of shrimp farming in Texas. Pp. 36-57 in Shrimp Culture in North America and the Caribbean, P.A. Sandifer, ed. Baton Rouge, La.: The World Aquaculture Society.

Chesapeake Bay Program. 1991. Report and recommendation of the Non-point Source Evaluation Panel, CPB/TRS 56/91. Annapolis, Md.

Chew, K.K. 1990. Global bivalve shellfish introductions. Journal of the World Aquaculture 21(3):9-22.

Colt, J.E., and D.A. Armstrong. 1981. Nitrogen toxicity to crustaceans, fish and mollusks. In Proceedings of the Bio-Engineering Symposium for Fish Culture,

L.J. Allen and E.C. Kinney, eds. Fish Culture Section, American Fisheries Society Bethesda, Md.

Courtenay, Jr., W.R. Regulation of aquatic invasives in the United States of America with emphasis on fishes. Unpublished manuscript.

Cross, S.F. 1990. Benthic impacts of salmon farming in British Columbia. Report to the British Columbia Ministry of the Environment, Water Management Branch, Victoria, B.C. 78 pp.

Daniels, H.V., and C.E. Boyd. 1989. Chemical budgets for polyethylene-lined, brackish water ponds. Journal of the World Aquaculture Society 20(2):53-60.

Dixon, I. 1986. Fish Farm Surveys in Shetland: Summary and Survey Reports, Vol. 1. A report to NCC, Shetland Islands Council and Shetland Salmon Farmers Assoc. FSC/OPRU/30/80. Orielton Field Center, Pembroke, Dyfed, Scotland.

Ellis, M. 1990. Decomposition processes on the pond bottom. Presented at Texas Aquaculture Conference, Corpus Christi, Tex. February.

Figueras, A.J. 1989. Mussel aquaculture in Spain and France. World Aquaculture 20(4):8-17.

Food and Agriculture Organization (FAO). 1977. Control of spread of major communicable fish diseases. Report of the FAO/OIE Government Consultation on an International Convention for the Control of the Spread of Major Communicable Fish Diseases. FAO Fisheries Reports, No. 192. FID/R192 (EN).

Gillespie, D. 1986. An inquiry into finfish aquaculture in British Columbia: Report and recommendations. Prepared for Government of British Columbia, December. 50 pp.

Gowen, R.J., and N.B. Bradbury. 1987. The ecological impact of salmon farming in coastal waters: A review. Oceanography and Mar. Biol. Annual Rev. 25:562-575.

Gowen, R.J., and D.S. McLusky. 1990. Investigation Into Benthic Enrichment, Hypernutrification and Eutrophication Associated With Mariculture in Scottish Coastal Waters. Summary of main report of Highlands and Islands Development Board, National Conservation Council and Scottish Salmon Growers Assoc., 13 pp.

Gowen, R.J., and H. Rosenthal. 1990. Presentations to the committee. Halifax, Nova Scotia, June 11-15.

Griffiths, R.W., D.W. Schlaesser, J.H. Leach, and W.P. Kovolak. 1991. Distribution and dispersal of the zebra mussel *(Dreissena polymorpha)* in the Great Lakes Region. Canadian Journal of Fisheries and Aquatic Sciences 48:1381-1388.

Herman, R.L., D. Collis, and G.L. Bullock. 1969. Oxytetracycline residues in different tissues of trout. Bur. Sport Fish. and Wildlife. Tech Paper No. 37. U.S. Department of the Interior. Washington, D.C.

Hetrick, J. 1991. Alaskan Aquaculture. Water Farming Journal 6(4):(April)10-13.

Hicks, B. 1989. Fish health regulations restrict industry not disease. Can. Aquaculture 6(1):27-28.

Hindar, K., N. Ryman, and F. Utter. 1991. Genetic effects of aquaculture on natural fish populations. Can. J. Fish. Aquat. Sci. 48:945-957.

Hopkins, J.S. 1991. Status and history of marine and freshwater shrimp farming in South Carolina and Florida. Pp. 17-35 in Shrimp Culture in North America and the Caribbean, P.A. Sandifer, ed. Baton, Rouge, La.: The World Aquaculture Society.

Hopkins, J.S., M.L. Baird, O.G. Grados, P.P. Maier, P.A. Sandifer, and A.D. Stokes. 1988a. Impact of intensive shrimp production on the culture pond ecosystem. Journal of the World Aquaculture Society 19 (1):37A (abstract).

Hopkins, J.S., M.L. Baird, O.G. Grados, P.P. Maier, P.A. Sandifer, and A.D. Stokes. 1988b. Impacts of Intensive Shrimp Culture Practices on the Culture Pond Ecology. Report from the Waddell Mariculture Center, of the South Carolina Marine Resources Division. Charleston, S.C.

Hopkins, J.S., A.D. Stokes, C.L. Browdy, and P.A. Sandifer. Submitted. The relationship between feeding rate, paddlewheel aeration rate and expected dawn dissolved oxygen in intensive shrimp ponds. Aquacultural Engineering.

Institute of Medicine (IOM). 1991. Seafood Safety. Washington, D.C.: National Academy Press.

International Council for Exploration of the Seas (ICES). 1984. Guidelines for implementing the ICES code of practice concerning introductions and transfers of marine species. Copenhagen: ICES Cooperative Res. Rep. 130, 1-20.

International Council for Exploration of the Seas (ICES). 1988a. Report of the ad hoc study group on environmental impacts of mariculture. Copenhagen: ICES Cooperative Res. Rep. 164, 83 pp.

International Council for Exploration of the Seas (ICES). 1988b. Report of the Working Group on Environmental Impacts of Mariculture. Hamburg, Germany, April.

Jacobsen, M.D. 1989. Withdrawal times of freshwater rainbow trout, *Salmo gairdneri*, after treatment with oxolinic acid, oxytetracycline and trimetoprim. J. Fish Diseases 12:29-36.

Jonas, M., J.B. Comer, and B.A. Cunha. 1984. Tetracyclines. In Antimicrobial Therapy, A.M. Ristuccia and B.A. Cunha, eds. New York: Raven Press.

Kalagayan, G., D. Godin, R. Kanna, G. Hagino, J. Sweeney, and J. Wyban. 1990. IHHN virus as an etiological factor in runt-deformity syndrome of juvenile *Penaeus vannamei* cultured in Hawaii. World Aquaculture Society, World Aquaculture 90, Abstract T17.2.

King, S.T., and J.R. Schrock. 1985. Controlled Wildlife Vol. III: State Wildlife Regulations. Lawrence, Kan.: Association of Systematics Collections.

Krom, M.D., C. Porter, and H. Gordin. 1985. Nutrient budget of a marine fish pond in Eilat, Israel. Aquaculture 51: 65-80.

Lightner, D.V. 1990. Viruses section: introductory remarks. Pp. 3-6 in Pathology in Marine Science. New York: Academic Press.

Lutz, R.A. 1985. Mussel aquaculture in the United States. Pp. 311-363 in Crustacean and Mollusk Aquaculture in the United States, J.V. Huner and E.E. Brown, eds. Westport, Conn.: AVI Publishing Co.

Manzi, J.J. 1985. Clam aquaculture. Pp. 275-310 in Crustacean and Mollusk Aquaculture in the United States, J.V. Huner and E.E. Brown, eds. Westport, Conn.: AVI Publishing Co.

Mathisen, O.A., and T. Gudjonsson. 1978. Salmon management and ocean ranching in Iceland. J. Agr. Res. Iceland 10(2):156-174.

McLaughlin, T.W. 1981. Hatchery effluent treatment—U.S. Fish and Wildlife Service. In Proceedings of the Bio-Engineering Symposium for Fish Culture, L.J. Allen and E.C. Kinney, eds. Fish Culture Section, American Fisheries Society, Bethesda, Md.

Mudrak, V.A. 1981. Guidelines for economical commercial fish hatchery wastewater treatment systems. In Proceedings of the Bio-Engineering Symposium for Fish Culture, L.J. Allen and E.C. Kinney, eds. American Fisheries Society, Bethesda, Md.

Munro, A.S.S., and I.F. Wadell. 1984. Furunculosis: Experience of its control in the sea water cage culture of Atlantic salmon in Scotland. ICES CM/F:32. 9 pp.

National Academy of Sciences (NAS). 1980. The Effects on Human Health of Subtherapeutic Use of Antimicrobials in Animal Feeds. Committee to Study the Subtherapeutic Antibiotic Use in Animal Feeds. Washington, D.C.: National Research Council.

Nehlsen, W., E. Williams, and J.A. Lichatowich. 1991. Pacific salmon at the crossroads: Stocks at risk from California, Oregon, Idaho, and Washington. Fisheries 16(2):4-21.

North Atlantic Salmon Conservation Organization (NASCO). 1990. Report on the Norwegian meeting. Loen, Norway, May.

Nose, T. 1985. Recent advances in aquaculture in Japan. Geojournal 10(3):261-276.

Paramatrix, Inc. 1990. Final programmatic environmental impact statement: Fish culture in floating net pens. Prepared for the Washington Department of Fisheries, Olympia, Wash.

Porter, C., M.D. Krom, M. Robbins, L. Brickell, and A. Davidson. 1987. Ammonia excretion and total N budget for gilthead seabream *(Sparus aurata)* and its effect on water quality conditions. Aquaculture 66: 287-297.

Pruder, G.D. 1991. Shrimp culture in North America and the Caribbean: Hawaii 1988. Pp. 58-69 in Shrimp Culture in North America and the Caribbean, P.A. Sandifer, ed. Baton Rouge, La.: World Aquaculture Society.

Richards, G.P. 1988. Microbial purification of shellfish: A review of depuration and relaying. J. Food Protection 51(3): 218-251.

Rosenberry, R. 1990. Shrimp farming in the Western Hemisphere. Presented at Aquatech 90, Malaysia, June.

Rosenfield, A., and F.G. Kern. 1979. Molluscan imports and the potential for introduction of disease organisms. Pp. 165-89 in Exotic Species in Mariculture, R. Mann, ed. Cambridge, Mass.: MIT Press.

Rosenthal, H. 1980. Implications of transplantations to aquaculture and ecosystems. Marine Fisheries Review 42(5):1-14.

Rosenthal, H. 1985. Constraints and perspectives in aquaculture development. GeoJournal 10(3):305-324.

Sandifer, P.A. 1988. Aquaculture in the West: A Perspective. Journal of the World Aquaculture Society 19(2):73-84.

Salser, B., L. Mahler, D. Lightner, J. Ure, D. Danald, C. Brand, N. Stamp, D. Moore, and B. Colvin. 1978. Controlled environment aquaculture of penaeids. In Drugs and Food From the Sea, Myth or Reality? P.M. Kaul and C.J. Sindermann, eds. Norman, Okla.: University of Oklahoma Press,

Sandifer, P.A., J.S. Hopkins, A.D. Stokes, and C. Browdy. Submitted. Preliminary comparisons of the native *(Penaeus setiferus)* and Pacific *(P. vannamei)* white shrimp for pond culture in South Carolina.

Sandifer, P.A., J.S. Hopkins, A.D. Stokes, and G.D. Pruder. 1991a. Technological advances in intensive pond culture of shrimp in the United States. Frontiers of Shrimp Research. Elsevier. New York, N.Y.

Sandifer, P.A., A.D. Stokes, and J.S. Hopkins. 1991b. Further intensification of pond shrimp culture in South Carolina. In Shrimp Culture in North America and the Caribbean, P.A. Sandifer, ed. World Aquaculture Society.

Sattaur, O. 1989. The threat of the well-bred salmon. New Scientist 29 (April):54-58.

Saunders, R. L. 1988. Algal catastrophe in Norway. World Aquaculture 19(3):11-12.

Seter, R.M. 1990. Potential within aquaculture issues of the use of genetic engineering and of the introduction of species. Unpublished manuscript. College of Marine Studies, University of Delaware.

Shumway, S.E. 1990. A review of the effects of algal blooms on shellfish and aquaculture. Journal of the World Aquaculture Society 21(2):65-104.

Sindermann, C.J. 1988. Disease problems created by introduced species. Pp. 394-98 in Disease Diagnosis and Control in North American Marine Aquaculture, 2nd ed., C.J. Sindermann and D.V. Lightner, eds. Amsterdam: Elsevier.

Thompson, A.G. 1990. The danger of exotic species. World Aquaculture 21(3):25-32.

Tibbs, J.F., R.A. Elston, R.W. Dickey, and A.M. Guarino. 1988. Studies on the accumulation of antibiotics in shellfish. Northwest Environmental Journal 5(1).

Toranzo, A.E., P. Combarro, M.L. Lemos, and J.L. Barja. 1984. Plasmid coding for transferable drug resistance in bacteria isolated from cultured rainbow trout. Applied and Environmental Microbiology 48:872-877.

Van Engel, W.A., W.A. Dillon, D. Zwerner, and D. Eldridge. 1966. *Loxothylacus panopei* (Cirripedia, Sacculinidae) an introduced parasite on a xanthid crab in Chesapeake Bay, U.S.A. Crustaceana 10:110-112.

Vermeij, G.J. 1991. When biotas meet. Understanding biotic interchange. Science 253:1099-1103.

Walton, J. R. 1988. Antibiotic resistance: An overview. Veterinary Record 122:247-251.

Waples, R.S., G.A. Winans, F.M. Utter, and C. Mahnken. 1990. Genetic approaches to the management of Pacific salmon. Fisheries 15(5):19-25.

Water Farming Journal. 1991. FDA adopts tough new policy on use of drugs in aquaculture. (September 28):4-6, 27.

Weston, D.P. 1986. The environmental effects of floating mariculture in Puget Sound. School of Oceanography, College of Ocean and Fishery Science, University of Washington, Seattle. 148 pp.

Weston, D.P. 1991. The effects of aquaculture on indigenous biota. Pp. 534-567 in Aquaculture and Water Quality, D.E. Brune and J.R. Tomasso, eds. Baton Rouge, La.: World Aquaculture Society.

Weston, D.P., and R.J. Gowen. 1988. Assessment and prediction of the effects of salmon net-pen farming on the benthic environment. Report to Washington Department of Fisheries, Olympia, Wash. 62 pp.

Whetstone, J.M., E.J. Olmi III, and P.A. Sandifer. 1988. Management of existing saltmarsh impoundments in South Carolina for shrimp aquaculture and its implications. Pp. 327-338 in The Ecology and Management of Wetlands, D.D. Hook et al., eds. Portland, Ore.: Timber Press.

Whiteley, A.H., and A. Johnstone. 1990. Additives to the environment of net-pen reared fish. Proc. Pacific Marine Fisheries Commission 42nd Annual Meeting, Seattle, Wash., October 16-18, 1989.

Wright, K. 1990. Bad news bacteria. Science 249:22-24.

Wyban, J.A., and J.N. Sweeney. 1991. Intensive Shrimp Production Technology; the Oceanic Institute Shrimp Manual. Hawaii: The Oceanic Institute. 158 pp.

Ziemann, D., G.D. Pruder, and J.K. Wang. 1990. Honolulu, Hawaii: University of Hawaii. Aquaculture Effluent Discharge Program, Center for Tropical and Subtropical Aquaculture, Year 1 Final Report.

5

Engineering and Research

ROLE OF SCIENCE AND TECHNOLOGY IN ADDRESSING MAJOR CONSTRAINTS

A broad range of economic, institutional, environmental, and social concerns can, to some extent, be addressed through advances in the science and technology base supporting marine aquaculture. Problem areas that are susceptible to mitigation through technological approaches include economic feasibility, market structures and product form, the regulatory framework for leasing and permitting, land and water use, ecological impacts, aesthetic issues, use conflicts, and public attitudes.

Summaries of the major issues follow, with examples of where science and technology can contribute to the resolution of related problems.

Economic Feasibility

Advances in technology can improve economic feasibility through (1) the creation of new capability, (2) the design of more productive (higher-yield) operations, and (3) the reduction of expenditures through more effective and efficient operations and the substitution of cost-effective capital investment for labor. Specific opportunities for improving marine aquaculture in these areas include:

- new culture systems that make possible the production of marine species in environmentally sound ways;
- improved technology for culture operations to utilize inputs more efficiently, increase productivity, and reduce costs of production and waste

disposal (e.g., water use and reuse, feeding technology, product inventory, product handling, waste disposal);

- technology that improves the cost-effectiveness of operations through intensification of culture systems, reduced operating costs, and increased productivity; and
- technology that reduces production uncertainty (e.g., through disease detection and treatment, inventory monitoring systems, and design of more seaworthy facilities), thereby reducing risk and the associated costs of capital, insurance, and other nonoperational factors.

Marketing and Product Information

Technology can enhance the quality and value of products in addition to increasing productivity and reducing costs. Examples are:

- harvest, transportation, processing, and packaging technologies that will allow aquaculture to deliver high-quality products in good condition to appropriate markets;
- technologies that can maintain high-quality standards and ensure wholesome and safe products; and
- new product forms for new and traditional aquaculture species.

Institutional and Regulatory Issues

Technology can be used effectively to address many institutional issues. Opportunities include:

- technology to diminish the amount of water or land necessary for culture and auxiliary systems, thus minimizing land/water use conflicts;
- information systems to improve communication with the public, provide relevant facts, make information more accessible, and generally increase understanding of the benefits and constraints of aquaculture;
- technology that will resolve issues associated with access to brood stock and seed/juvenile production from wild populations through achieving controlled reproduction, an understanding of improved nutritional requirements, and better knowledge of species life cycles;
- technology to better identify and control disease-related problems; and
- technology for the identification of cultured fish in order to differentiate among stocks for marketing and management purposes.

Environmental Issues

Marine aquaculturists must be sensitive to issues of common resource use and must seek ways to reduce pollution and other environmental impacts. Science and technology can contribute significantly to this goal by

• achieving waste treatment and removal, and water and feed delivery, that alleviate pollution and discharge problems in culture and auxiliary systems;

• providing means to minimize disease transmission in culture operations and thereby improve disease prevention and management;

• providing improved culture and auxiliary systems (for open ocean production, closed systems, and ocean ranching) that mitigate the ecological impacts;

• providing alternative, nearshore, culture systems that can mitigate conflicts with recreational, commercial, and navigational use;

• providing innovative culture systems that address the aesthetic issues associated with nearshore operations (i.e., by use of submerged cages, offshore production, closed systems, and ocean ranching);

• developing analytical techniques and computer models to simulate the environmental impact of aquaculture operations (Brune, 1990);

• improving stock sterilization capability that prevents reproduction in cultured animals and prevents genetic dilution of wild stocks from escaped fish;

• improving harvest, packaging, and transportation systems to alleviate potential sanitation and public health concerns; and

• providing the capability to identify genes that control growth (a capability that has been achieved with nonfish food species).

Socioeconomic Issues

The development of technology for marine aquaculture not only can improve the economic situation for producers but can contribute to the year-round economic health of rural communities as well. Specific examples include (1) providing employment for laborers who work on aquaculture farms, and (2) creating or augmenting the need for suppliers and processors that, in turn, provide employment.

INTERDISCIPLINARY SYSTEMS DESIGN

Marine aquaculture systems require individual elements designed so that each can function effectively alone and can also function in concert with other elements to comprise an interactive system. For example, a simple home aquarium may be viewed as a system made up of a few common elements—a tank, air pump, air diffuser, water pump, and filter. Aquaculture systems, although conceptually similar, are much more complex in terms of design, operation, and management. The biological functions of the fish must be taken into account, including special requirements associated with intensive culture operations. Consequently, the design of a commercially viable system involves considerations beyond purely engineering criteria for integrating the elements into a working physical system (Huguenin and Colt, 1989).

Design, operation, and management are further complicated by the need for profitability, the risks and challenges associated with the intensive production of animals, and the necessity of working in a frequently hostile environment—the ocean. The project team must select an adequate site; establish the physical, chemical, and biological requirements for the species in culture; and also design a system that is economically viable. An interdisciplinary approach is needed to achieve all these objectives. The engineer, the biologist, and the entrepreneur must collaborate effectively in order to solve problems and develop improved technology for marine aquaculture, an arrangement not easily achieved in an era of increasing specialization.

Although technology development is needed for the commercial success of marine aquaculture, research on the biology of potential cultivars is also essential. One of the principal constraints to economic viability is the lack of sufficient biological information necessary as design criteria for fish culture. Too little is known about life cycles, the means of controlling reproduction, the environmental and nutritional requirements of larvae, the causes and effects of stress, and biological and environmental requirements in general. Effective interdisciplinary systems design can be realized only if the biological criteria for design are well understood.

Following are discussions of the major areas in which interdisciplinary research and developments can make significant contributions to the advancement of marine aquaculture and to the resolution of many outstanding issues that presently constrain the industry. First, auxiliary systems that are an essential part of all types of culture systems are discussed. Then culture and confinement systems are discussed in the context of those that are adaptable to nearshore locations, those that can be used onshore, and systems compatible with offshore production.

Auxiliary Systems for Fish Culture

Improvement and development of the various auxilliary systems that are required for culturing fish are essential to the establishment of commercially viable marine aquaculture. Aquaculture systems must ensure the confinement or physical support necessary to hold the animal, as well as provide the auxiliary elements required for healthy aquatic life (Fridley et al., 1988). Key needs are adequate water with adequate oxygen, effective feed and feeding systems for marine species, waste treatment, and sensors and monitoring capability. Expert systems, including computer monitoring and prediction capability, can be very helpful as well. Most of these needs are provided by auxiliary systems and are basic to the cultivation or husbandry of any animal, terrestrial or aquatic.

Hatchery Systems

The culture of most species requires a hatchery in which to collect, incubate, and hatch eggs and/or rear larval fish and young juveniles. Hatcheries require rigorous controls and careful management. The young animals are intolerant of adverse water temperature and quality, and often are difficult to feed. A variety of jars, racks, sacks, and other containers have been developed to hatch eggs and to set the spat of shellfish. Special diets and

Hatchery tank with a Macdonald jar—an incubation container that provides an environment conducive to egg development with minimum stress and minimum opportunity for disease.

special ways of presenting the feed have been created. Each species tends to have some unique requirements that lead to continual innovation as advances are made with current species and as new species are cultured.

Hatchery development can be a limiting factor in attempts to culture new species. Hatchery limitations generally tend to be more biological than technological. The intensive practices (high population density) of hatcheries and the relatively short time that animals are in the hatchery generally result in lower water requirements and smaller facilities than for the grow-out stage of development. This smaller scale of operations tends to limit the level of environmental and public concern. However, in the future, the pursuit of offshore systems may present technology problems related to the design of offshore hatcheries or to the transport of juveniles from an onshore hatchery to an offshore culture facility. In any case, the biological information needed to produce high-quality stock consistently and economically is often a limiting factor in achieving cost-effective hatchery production.

Feed and Feeding Systems

The feeding habits and the morphology and composition of feed vary greatly by species. Consequently, different artificial diets and feeding systems need to be developed in each kind of culture operation. A large body of information is available on feeds and feeding systems for salmonids and catfish (NRC, 1974a,b, 1977; Halver, 1988; Lovell, 1989). Considerable information is also available regarding the nutritional and feeding requirements of oysters and lobsters (Conklin et al., 1983). Future efforts should build on existing knowledge and focus on the special needs of different marine species. Of particular importance are nutritional requirements, effective feeding systems, improved efficiency of feed utilization, and alternative protein sources, especially in relation to protein quality and specific requirements during different periods of the life cycle.

The larval and juvenile stages of many marine species are relatively small—perhaps 2-3 millimeters (mm) at the time initial feeding is required. This factor presents unique problems with regard to the size of food offered, the acceptability of prepared food versus live food and the delivery system (Bromley and Sykes, 1985; Holt, 1990, 1992). Microencapsulated diets have been under development to replace live feeds for larval and juvenile stages, but they are not yet entirely sufficient (Kanazawa et al., 1989). Research on better attractants to promote feeding or on improved feed palatability should lead to lower feed conversion ratios (weight of feed consumed to weight of fish produced—generally between 1 and 3).

Nutritional requirements of a given species change with the transitions from larval to juvenile to adult stages. Nutritional requirements need to be better defined for each species and for each life history stage so that rations can be tailored to meet the precise dietary requirements of the species and stage (Ratafia and Purinton, 1989). In the future, rations will be tailored not only to the requirements of the species under culture but also to the characteristics of the culture systems (e.g., pond system, water reuse system).

Protein is the single most expensive and essential component of fish feeds. Consequently, the substitution of less expensive sources of protein for fish and other animal meals in feed could substantially reduce production costs. Use of soybean meal to replace animal protein has been moderately successful with some species (Cowery et al., 1971; Cho et al., 1974). Other researchers have used poultry egg proteins (Davis et al., 1976; Conrad et al., 1988) or nematodes (Biedenback et al., 1989) to replace fish protein.

Researchers have investigated a number of feed additives, including antibiotics and other medications (Strasdine and McBride, 1979; Marking et al., 1988); vaccines (McClean and Ash, 1990); growth hormones (for review, see Donaldson et al., 1978), drugs to increase metabolic efficiency (Santulli et al., 1990); and synthetic reproductive hormones (Yamazaki, 1983). Feed formulations are being developed to provide natural or synthetic pigments (Yamada et al., 1989) and to deliver stable and water-insoluble forms of necessary vitamins (Shigueno and Itoh, 1988; Grant et al., 1989).

Because feed can release large amounts of nitrogen and phosphorus, and thus cause localized eutrophication in some areas, improved feeds could mitigate concerns about eutrophication. Ketola and his associates have investigated the problem of phosphorous enrichment of receiving waters via salmon feeds and the effects of feed improvements in reducing such releases (Ketola, 1975, 1982, 1985, 1988, 1990; Ketola et al., 1985, 1990). Feeds that result in more efficient assimilation of nutrients are needed to reduce the waste treatment requirements and limit environmental impacts.

Consideration should be given to the design of feeds that, if uneaten, can contribute to other links in the food chain. Waste products from feeds, for example, could serve as a primary source of nutrition in a serial polyculture system (i.e., in which water and nutrients pass from one containment vessel with one species to another vessel containing a different species) (Wang, 1988; 1990).

The feasibility of altering the nutritional value of aquaculture products for humans or of enhancing other components to improve the marketability or palatability of farmed aquatic products is also under investigation. Assessments of the relative fatty acid profiles of farmed and wild fish are already under way, partly as a result of interest in nutritional information (Nettleton, 1990). This information will serve as a guide to the development of "finishing diets" that will provide consumer-ready products with the most nutritionally healthful compositions possible.

The diversity of feeds—pellets, algae, seaweed, small and large—required for different species in culture creates the need for a diversity of feeding systems. Feeding systems in need of development include systems for increasing the efficiency of utilization of the nutrient, decreasing waste production (in terms of feed that is not consumed and feces production of the culture species), delivering micronutrients and medications, and promoting by-product usage. The development of feeds and feeding systems that can provide feed at a rate consistent with the ability of the fish to consume it would enhance the cost-effectiveness of all feeding systems.

Such systems would also provide environmental benefits from reduced waste and water pollution in both the rearing and the effluent receiving waters, including reduced release of additives such as antibiotics. Design parameters that need to be understood include presentation of the food, frequency and rate of feeding, physical properties of feed particles, and impact of feeds and feeding methodology on wastage, growth, feed utilization, and predator species. For example, broadcasting feed over the water surface for juvenile finfish can be advantageous in getting the feed to the fish, but the presence of the fish at the water surface may attract bird predators.

Broadcast feeding of shrimp in lined seawater ponds.

Waste Treatment Systems

Treatment of wastes must be an integrated part of water reuse systems (discussed later in this chapter) and also may be required in flow-through and cage systems (Alabaster, 1982). Water disinfection and removal of solid (excess feed and fecal material) and dissolved (ammonia and dissolved organics) wastes are essential in any onshore water reuse system. In most cases, proper site selection for onshore or nearshore systems can minimize problems associated with waste. Dispersal or dilution of wastes for cage culture can be facilitated by proper site selection, but mechanical means of dispersing or treating wastes and filtering effluents are needed for some situations. A fanlike pumping systems placed below cages reportedly can flush large quantities of water through the system (Aase, 1985). In other cases, collection of wastes is required. Waste collection systems vary greatly for different culture systems. For intensive culture in ponds and tanks, solid waste collection sometimes can be accomplished with the simple addition of a settling tank or pond. However, more cost-effective methods of waste collection and dispersal need to be developed.

Reuse systems employ a wide variety of treatments to achieve the desired water quality changes. These may include the following components: filters, screens, clarifiers, oxygen injection, aeration, biofilters for dissolved organics and ammonia removal, chemical ammonia removal, heat exchangers, ultraviolet light disinfection, ozone disinfection, and chlorine disinfection (Miller and Libey, 1985; Malone and Burden, 1988). Biofilters are a critical component in the development of commercially viable recirculating systems, and research in this area continues to be very active (e.g., Brune and Piedrahita, 1983; Kruner and Rosenthal, 1983; Miller and Libey, 1985; Rogers and Klemetson, 1985; Malone and Burden, 1988; and Kaiser and Wheaton, 1991).

Dead and diseased organisms present another waste disposal issue faced by marine aquaculturists. Management of this waste may be significantly different from that of fish processing plants because the risk of disease transmission to other cultured fish and to wild fish must be minimized in aquaculture operations. However, it is also essential that processing plants and other facilities take the steps necessary to ensure that diseases are not transferred to wild populations. Clearly, both commercial fish processing facilities and aquaculture processing facilities have to dispose of animal wastes. The technical issue of disposal can be accomplished by utilizing current land-based disposal methods including landfills or incineration. However, the continued use of landfills and incineration in the future may be problematic because of limits on their availability or environmental concerns. Alternative means of disposal need to be developed.

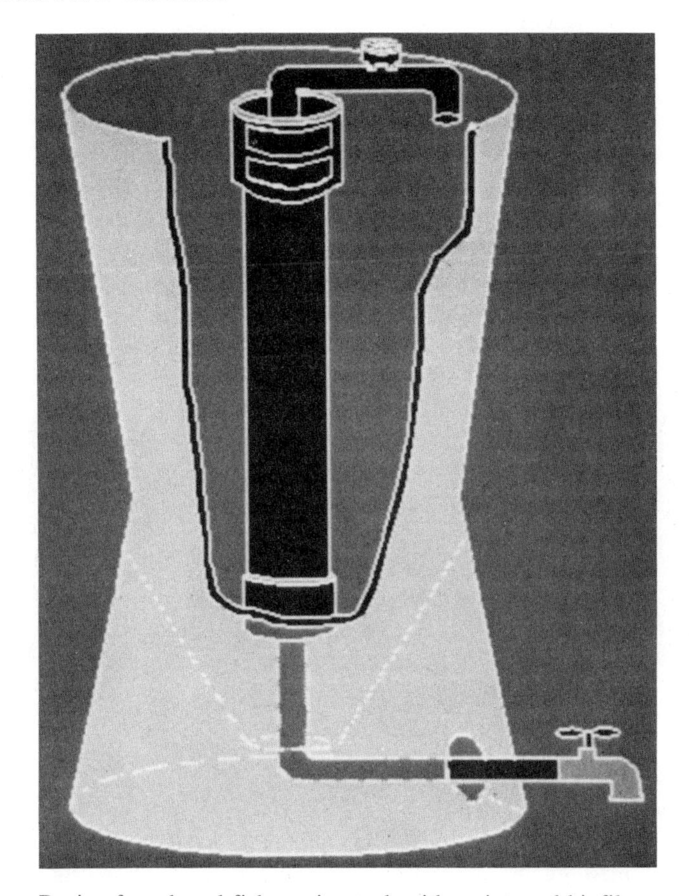

Design for a larval fish-rearing tank with an internal biofilter.

Sensors and Monitoring Systems

A sensor and monitoring system can provide valuable information and thereby improve the chances of success for marine aquaculture. For example, oxygen levels fluctuate in response to different internal or external factors, and these variations can stress or even kill the animals if adequate aeration is not provided. When fluctuations are not fatal, unsatisfactory fish health and growth, inefficient feed utilization, and poor reproduction can result (Wyban and Antill, 1989). Oxygen concentrations in ponds are particularly troublesome and difficult to measure and predict (Losordo et al., 1988; Piedrahita, 1991). Seemingly identical ponds within a single farm often have different oxygen conditions. Oxygen levels are changing constantly and can vary significantly even in the same pond.

Paddlewheel aeration of an earthen aquaculture pond.

Accurate and reliable sensors to monitor basic water quality parameters in seawater are not presently available. Existing automatic systems for continual in situ oxygen measurements are costly to install, require frequent and skilled maintenance, and typically have a short operating life. The marine environment causes rapid deterioration of equipment; metabolic by-products and other impurities in seawater interfere with the measurement process; and the cost for the multitude of measuring points needed is high.

Oxygen is just one of many parameters that are currently difficult to monitor and control with available instrumentation (Kaiser and Wheaton, 1991). Others parameters of special significance and technical challenge are ammonia, carbon dioxide, pH, salinity, light transmission, and biomass. Even when measurements are not especially complex technically—such as the determination of flow, water level, and temperature—existing equipment is subject to biofouling and corrosion. Improved instrumentation and automatic monitoring systems are needed to solve these problems.

Expert Systems

The widespread availability of relatively inexpensive computers, together with the development of improved sensors and monitoring equipment, is accelerating the evolution of more advanced monitoring, control, and prediction systems, collectively referred to as expert systems. A simple expert system for aquaculture would monitor biomass of fish, temperature, and concentration of oxygen. As the temperature or biomass changes, the computer would calculate the appropriate feed amounts and command the mechanical feeder to release the desired amount, provided the oxygen concentration was adequate for good utilization of the feed. The expert system maintains the amount of food offered at a preprogrammed level, but avoids feeding in the event that the oxygen concentration has been lowered, and thereby reduces potential waste, increases food conversion efficiency, and maximizes growth, given the particular growing environment. With the addition of yet another computer routine, this expert system would determine whether changes in the oxygen level after feeding match expected values. If not, the presumption would be that the feeders had not operated properly or that the fish are not consuming the feed as expected, and the operator would be signaled to take action.

The characteristics of an expert system provide the ability to sense conditions, to take a variety of control actions in response to conditions that increase costs, to record events, to detect possible system failures, and to notify operators or set an alarm. The evolution of reliable expert systems specific to fish culture could have a major impact on the cost and risk of operations, and would benefit environmental studies and marketing control (Palmer, 1989; Weaver, 1990).

Production Grow-out Systems

Facilities for culturing marine species can be categorized as *nearshore* (located in coastal waters very close to the shore, i.e., within easy sight of other coastal users); *onshore* (located on land near the ocean or estuaries where seawater can be pumped to the facility); and *offshore* (located in the ocean at least somewhat away from the shore, i.e., in the vicinity of offshore oil rigs or independently in deep water). A long-term objective for marine aquaculture that offers a variety of potential advantages is to locate projects inland or offshore, away from the socially and environmentally sensitive nearshore area.

Nearshore Systems

Until new technology makes relocation logistically and economically feasible, it will be necessary for most marine aquaculture operations to be located in nearshore shallow bays and estuaries. Therefore, high priority should be given to technological developments that will make nearshore aquaculture less objectionable and intrusive to other competing uses of the coastal zone, aesthetically more acceptable, and environmentally benign if not beneficial.

Mollusks are grown in the nearshore either on the shallow bottom (bottom cages) or on one of a variety of support systems (racks, bags, suspended lines, stakes). Bottom culture of mollusks is basically the culture of shellfish in their natural growing habitat. Seed are dispersed over a coastal or estuarine bottom area that is a productive environment for the species being sown. The animals are grown in a relatively natural and nonintensively managed setting; the culturist harvests mature animals much as one would gather animals from the wild. However, a variety of support systems have been created to intensify production and increase its efficiency. Support systems, if well conceived, can reduce labor requirements, increase productivity, and increase the yield from seed. Commercial support systems include structural racks and trays on which shellfish grow, supported bags that confine groups of shellfish, and suspended lines on which shellfish can grow. In each case, the support system provides a form of three-dimensional, off-bottom support, and the physical device used for support also facilitates harvest.

Typical nearshore culture systems for finfish are cages or net pens mentioned for salmon in Chapter 2. Cages or net pens[1] are usually supported by a floating structure. Each cage or pen is surrounded by netting or a similar mesh material with a bottom about 1 to 5 meters (m) below the water surface. Water flow and flushing are provided by natural currents and tidal flow. Biofouling is a common problem. At a minimum, cages must be

Cage facility for culturing Atlantic salmon in Hitra, Norway. Both circular and rectangular cages can be seen.

Experimental stake culture of oysters *(Crassostrea virginica)* in a shrimp pond at the James M. Waddell, Jr., Mariculture Center in Bluffton, South Carolina.

cleaned following harvest prior to starting a new cohort. Typically, walkways along the periphery of the cages provide worker access for feeding, inspecting, removal of any dead fish, maintenance, and harvest. Structures that support several cages usually provide space for feed storage and access to equipment as well. Limited technical studies have been published on cage systems (see Beveridge, 1987; Kerr et al., 1980; Linfoot and Hall, 1987).

Although technical and economic feasibility has been established for the nearshore culture of some marine species, commercial viability is challenged by numerous environmental, institutional, and social issues that increase costs. New technological advances are needed to permit aquaculture to flourish in the nearshore environment, which is often exploited for other uses or degraded and polluted from intensive development. Other users of the nearshore often oppose surface structures needed for marine aquaculture and thereby curtail efficient culturing of some species. For example, suspended off-bottom culture of mollusks is generally far more productive and successful than bottom culture and is widely practiced in most parts of the world; however, suspending such cultures from rafts or buoyed long lines is generally considered to be unacceptable in most U.S. coastal waters. Similarly, the culture of seaweed on rafts, of pelagic finfish in floating cages, and of mollusks in pens encounters opposition.

Improved design and location of such surface structures or designs for their deployment beneath the sea surface would make them less objectionable aesthetically and decrease the potential for interference with boaters by permitting passage of small boats through, around, or over the structures. An innovative approach to containment and harvesting of finfish is the use of sonar or electrical fencing as suggested by Balchen (1987) for use on salmon. Research is needed to determine the feasibility of such techniques.

Opportunity for improvement exists also in the bottom culture of bivalve mollusks—oysters, clams, scallops, mussels (Korringa, 1976; Imai, 1977)—which is one of the least objectionable forms of aquaculture with respect to visual impact and interference with other uses of the coastal zone. Improved methods of planting and harvesting stocks are needed to minimize impact on the benthic ecosystem and decrease the resuspension of bottom material. The carrying capacity of various environments for cultured mollusks needs to be determined.

Pollution by waste from unused feed and feces is of particular concern with floating cage operations located nearshore (Weston, 1986). Improved technology would aid in site selection and in the operation of cages to avoid conditions under which accumulation of waste will occur. Once cages have been properly sited, technology can provide the capability to control the rates and intervals of feeding to avoid overfeeding. The ability to monitor accurately the biomass of fish in individual cages would enable feeding the amounts required for a reasonable growth rate with minimal waste and pollution.

Human fecal contamination of coastal waters and the resulting risks and dangers to public health are major constraints to shellfish farming. The traditional methods of detecting contamination are inadequate and result in many areas of coastal waters being unnecessarily removed from production. New techniques are needed involving the use of better indicators of human pollution, preferably direct monitoring of the pathogens themselves.

Shellfish do not become infected with human pathogens but accumulate the microorganisms in their intestinal tracts and become incidental temporary carriers. With proper technology, shellfish can be quickly and effectively cleansed of harmful bacteria and made safe for human consumption through a process known as depuration (Richards, 1988). Depuration is a natural self-cleansing of shellfish made possible by their biological need to ingest and discharge water to expel waste. Simply described, the shellfish are placed in clean water where they pump water through, and thereby purify, their bodies. Much more effective, dependable, and economical depuration systems are required before the practice will be widely accepted and adopted. Also, the efficacy of depuration in purging animals of viruses on a commercial scale has not been evaluated.

Shellfish may also become contaminated with dangerous poisons through ingestion of toxic microorganisms from the water. An increase in the incidence of phytoplankton blooms (i.e., red tides) of such toxic algae has been observed around the world. These events are sometimes correlated with outbreaks of paralytic shellfish poisoning (PSP) and diarrhetic shellfish poisoning (DSP) in humans, when people ingest shellfish contaminated with such toxins. Current methods for toxicant detection are often based on bioassays that are slow and expensive to conduct. As a result, harvest closures frequently are implemented on a seasonal basis rather than as a response to the actual detection of toxic conditions. Simpler, quicker, and more dependable methods are needed for detecting the presence of such toxins, along with methods for detoxifying or otherwise depurating the contaminated mollusks.

Toxic algal blooms may prove dangerous or lethal to the cultivated animals, particularly to cage- or pen-cultivated finfish. Mobile culture systems could be developed that can be raised or lowered within the water column (i.e., below a toxic algal bloom that might be concentrated at or near the surface) or towed away from a local concentration of pollutants as a crop-saving measure.

Onshore Systems

The development of onshore systems would enable the movement of culture operations inland from the nearshore coastal waters where many other competing activities tend to take place such as recreational boating and fishing. The major factors limiting further expansion of the industry would then become technical and economic rather than political and institutional.

Onshore culture systems are based on fixed rearing units of various types, such as ponds, tanks, and raceways. Ponds are shallow (typically 1 to 1.5 m deep) reservoirs of water. Historically, large earthen or lined ponds are most common (Lannan et al., 1985). The water supply to the ponds is often intermittent, although continuous supplies are also used. Water supply overturn rates (time to replace pond water) typically are measured in days or weeks, and aeration is often required intermittently. Tanks are shallow (typically 1 to 1.5 m deep), aboveground, circular structures. Raceways are elongated concrete or fiberglass channels. Tanks and raceways most often are constructed of concrete or fiberglass. The water overturn rates typically are measured in minutes or hours. Thus the biomass loading (in kilograms per cubic meter or kilograms per square meter) in tanks or raceways is generally much higher than in ponds; biomass loadings may reach 20 percent or more of the water volume. Aeration or oxygenation is usually required.

Use of pond systems for marine aquaculture is limited primarily by the availability of suitable sites (including the large amount of space required) and environmental concerns. Tanks and raceways are suggested as inten-

Intensive tank facility that incorporates water recirculation and the use of liquid oxygen (AquaFuture, Paso Robles, California). Water is introduced into the tanks by a radial pipe that directs water tangentially into the tank to provide circulation.

sive alternatives to pond culture to reduce space requirements and provide controlled environmental conditions, especially temperature (Arnold et al., 1990). Environmental issues of water use and effluent discharge that are associated with pond systems also exist with tanks and raceways. The preferred system depends largely on land cost, the cost and availability of labor, and the cost of capital. Ponds are best under conditions of low land and labor costs. Tanks and raceways require less land and labor but more capital facilities and equipment. Tanks, and to a lesser degree raceways, also can be employed with water reuse systems (to be discussed later) and therefore are preferred where water is expensive or in short supply, and where environmental concerns about water supply or discharge can be addressed by their use.

Site selection is a key component for the success of any aquaculture operation, including land-based systems. The further development of onshore systems awaits the determination of particular marine species that can be cultured in enclosed and crowded conditions. High-density recirculating systems have been tested for animals such as red drum, freshwater prawns, tilapia, and penaeid shrimp (Reed, 1989). At the present time, red drum, shrimp, striped bass, sturgeon, salmon, and abalone are being cultured onshore, at least on a limited scale. As production levels intensify, oxygenation and aeration systems become components of the culture system. All species require sufficient quantities of oxygen for survival and growth. Cost-

Construction of a greenhouse for the intensive closed-system culture of shrimp in a raceway.

effective aeration systems for both daily and emergency use are available for transfer from other fields.

Water Supply Systems Marine animals typically require large quantities of high-quality seawater. The water supply systems of marine aquaculture projects, therefore, are critical to the success of the operation and are an important component of the capital and operating costs. Several approaches are possible:

- open flow-through systems;
- systems that recirculate and thereby reuse some of the water; and
- closed systems (only makeup water is added to compensate for evaporation and leakage).

Flow-through systems use the largest water quantities for a given production level. They require a high level of reliability for the pumping system and a discharge capacity equal to the flow rate. Concomitant economic implications of flow-through systems for marine aquaculture include the following:

- the cost of water is high because the entire flow must be pumped and, if needed, must be heated, cooled, or treated;
- the total cost is related to the volume of water required and the height of lift; and

- flow-through systems require continuous flow, and system reliability is more critical than for intermittently operating systems that pump at high tide only.

Water cost is an important factor in either freshwater rearing (e.g., anadromous fish nurseries and rearing) or systems using pumped seawater. Most onshore systems, therefore, can benefit from some form of water reuse that either reduces energy and facility requirements for pumping and temperature control or reduces freshwater consumption where it must be purchased or pumped. Although a broad definition of reuse systems can apply to pond, raceway, or tank systems, reuse systems most typically refer to tanks. For the purposes of this discussion, the term "reuse system" is defined as a

Harvesting shrimp in a greenhouse with raceway-recirculating system.

system that recirculates all or part of the water passing through it using one or more processes to improve its quality. Aquaculture water reuse encompasses a variety of system designs ranging from relatively simple aeration systems with limited recirculation to 100 percent water recirculation for complete environmental control (Losordo, 1991). The need for developmental research in reuse systems is well accepted, but the technology is presently in operation in a number of noncommercial systems in the United States. Two examples are the Dworshak steelhead hatchery (Idaho) and the Living Seas Aquarium exhibit at Epcot Center (Florida).

The Dworshak hatchery was constructed in 1966-1967 in an attempt to mitigate damage to natural runs of the anadromous rainbow trout (the steelhead) from the construction of dams on their breeding rivers. Initially, 25 rearing tanks were connected to a 15,000-gallon-per-minute (GPM) reuse system that provided treatment of water as it left the tanks and partial reuse in these same tanks (Carey and Kramer, 1966). The primary justification for employing a reuse system at Dworshak was to reduce the energy and capital costs associated with cooling, heating, and disinfecting the water supply. The original filter system has long since been replaced by other systems, which, in turn, have also been replaced. Today, three systems are in place with a total flow of 45,000 GPM, and although research continues, technology development is mature for this application. The Dworshak hatchery has served as a center of reuse research and development for the federal government.

Most of the major marine aquariums of the world are based on closed reuse systems—little water is added to the process, even though the biomass of fish and marine mammals may be significant. For example, the major tank in the Living Seas Aquarium exhibit at Epcot Center in Florida holds 6 million gallons of water. The contents of the tank are recirculated every four hours at a rate of 25,000 GPM, and the daily inflow is 1,500 gallons. In many ways, this type of closed system is technically feasible for marine aquaculture operations. However, a combination of technical and economic factors constrains the development of these reuse systems for marine aquaculture applications. The economic situations of a food producer, a recreational aquarium, and a fish hatchery are very different. Although technically successful reuse systems are in service, cost-effective technologies applicable to commercial aquaculture still need to be developed (Mayo, 1989).

Offshore Systems

Use of offshore or "open-ocean" production systems could alleviate many of the institutional, regulatory, and environmental problems associated with coastal marine aquaculture activities by moving them out of the sensitive region of the coastal zone. Achieving such a transformation, however, is dependent on innovations in technology and design that will create economical offshore systems. Furthermore, the issue of the appropriate private use of public waters would not be circumvented by going offshore. Also, it must be kept in mind that a regulatory framework for such operations has yet to be established (see discussion in Chapter 3).

The most crucial factors constraining further development of marine aquaculture offshore are the capital and operating costs, the safety and efficacy of workers in the offshore environment, and the reliability of systems in the absence of continuous monitoring by personnel. Currently, capital costs would appear to prohibit profitability under most offshore conditions. The systems and materials available are expensive and have limited life in the harsh offshore marine environment. Efforts focused on the design of innovative culture systems and the development of low-cost durable materials are essential prerequisites to the economic feasibility of offshore operations. For example, cost-effective materials for marine use are critical to long-term maintenance, increased operating life, and reduced risk. Also essential are the design of either rigid or flexible structural systems, means of confinement, and anchoring systems that can withstand the physical forces of the offshore environment while still protecting the animals. Other critical engineering considerations are animal behavior characteristics related to containment and the effects of wave action (Gowen, 1988).

Interest in offshore marine aquaculture has been increasing recently. This interest is evident from two international meetings that have addressed engineering for offshore systems. The first meeting was held in October 1990 in Glasgow, Scotland (Institute of Civil Engineers, 1990). That meeting focused on the engineering problems associated with moving aquaculture operations (predominantly cage culture) from highly protected locations in fjords and estuaries to more exposed sites in the near coastal zone. The second meeting, a workshop held in September 1991 at the East-West Center in Hawaii (NSF, 1991), brought together biologists, engineers, researchers, and practitioners to explore the opportunities of offshore aquaculture, with particular attention to moving aquaculture operations even further away from protected coastal waters. Although technical problems and economic feasibility present major challenges to offshore operations, there is increasing optimism, based on experience, that aquaculture can be successful in the offshore environment (NSF, 1991).

Recently, interest has been expressed in the possibility of stabilizing atmospheric carbon dioxide (a major contributing factor to projected changes in global climate) through huge open ocean farming of macroalgae (the area needed to absorb significant amounts of atmospheric CO_2 is postulated to be slightly more than half the size of the contiguous 48 states) (EPRI, 1990).

Considerable research and technology development will be needed before offshore aquaculture can be commercially successful. The systems dynamics of cages need to be understood so that designs can provide (1) the functional requirements of the species being reared, (2) adequate equipment life and reliability, and (3) economical performance. Innovative confinement systems—including nonmechanical means such as electrical, sound, and light behavior conditioning—need to be developed for finfish along with new approaches to providing support media for shellfish. A broader knowledge base is necessary to facilitate better site selection, determined by considerations of water currents, environmental impacts, and anchoring or mooring requirements. Model and field testing is needed to evaluate component and systems design. New and improved construction materials and anchoring/mooring systems need to be developed. The potential for artificial upwelling to provide enhanced nutrition in select locations must be evaluated. Underlying these technical and engineering advances, an expanded base of biological research is necessary to provide the requisite knowledge of the life history of candidate species, nutrition requirements, fish behavior, controlled reproduction, and ecological impacts.

Existing Offshore Structures The use of existing offshore structures could alleviate some of the excessive cost of offshore operations. For example, structures that are built for other purposes, such as offshore oil platforms, can be utilized for shellfish production. Shellfish, clams, and mussels grow in large numbers when attached to such structures. An additional advantage is that these structures are privately owned, and therefore harvest rights can be controlled. One such example is a private firm using oil platforms off the coast of California. Meek (1990) reported that although the Food and Drug Administration (FDA) was concerned about potential sanitation problems from culture in conjunction with oil and gas platforms, studies of the levels of numerous contaminants yielded no evidence of a problem. Finfish cage culture also can be moored on existing offshore structures. In the Gulf of Mexico, experimental production of red drum in cages is based on an offshore oil platform. Existing structures that are available for this kind of multiple use or reuse are limited, of course.

Offshore Cage Systems Use of net pens for offshore production could relieve some of the siting concerns associated with net pens in coastal waters but raises a number of technology-related questions. Functionally, offshore cage systems are very similar to nearshore net pens, but they are much more costly with presently available technology. The concept also introduces a whole new range of technical and engineering problems that have to be addressed if offshore cages are to become commercially feasible. An offshore cage must be wave transparent, be designed to withstand the significant energy of the unprotected environment offshore or to be moved from place to place to avoid heavy seas, or be under water. Forster (1990) noted that current models are mostly of the transparent type and cost more than two times the amount of traditional nearshore cages.

Innovative research has been under way in California to develop a flexible cage system that enhances water exchange through what has been called "hydrodynamic mariculture" for offshore, in-the-sea cultivation of abalone and macroalgae. By following examples of the successes of agronomic researchers and agricultural engineers, this bioengineering approach to agriculture could also be applied to U.S. marine aquaculture (Neushal, 1990).

Several cages have been developed internationally for offshore use. No less than 18 trademarked systems for confined rearing of finfish were mentioned in a recent workshop (NSF, 1991), some of which were reported to be under development for offshore applications and others in commercial operation. A variety of sizes, shapes, and constructions have been used. For example, one type, known as the Japanese "Bridgestone" cage, uses 10- to 16-m-long, 0.4-m-diameter pressurized rubber cylinders to form a flexible collar from which a net is suspended (Beveridge, 1987). The collars may contain 6 to 10 sections, resulting in cages that are up to 50 m in diameter. A French company reported on a vessel-based operation anchored 3 kilometers (km) from the shore of Monaco (NSF, 1991). Hatchery and juvenile rearing were reported to be conducted on board the vessel, and grow-out was completed in cages in a semiexposed location near Malta. Surplus oil equipment was used in the system to keep capital costs to an affordable minimum. An Irish firm operated a system described as being for open-ocean use. It consisted of 12 cages, 14 m^2 and 15 m deep, suspended in a rigid framework that provided workways and flotation. The unit featured a computerized feeding system. This system was moored by two anchors in the shelter of an island and has not yet been tried in a truly open-ocean environment (Flynn, 1990). The costs of building and operating the structure were high.

Designs for fixed structures that move with the waves (i.e., up and down on a system of poles) or are submerged beneath the waves (Clarke and Beveridge, 1989) have been tested. A Canadian firm has tested a German spherical cage design (the Kiel cage) and is developing a design that can be fabricated in sizes from 12 to 30 m in diameter (NSF, 1991). The cage can be submerged to avoid heavy seas, and thereby reduce both mechanical stress and stress to fish, and to reduce interference with others. The spherical shape allows easy rotation for repair, maintenance, and harvesting by use of a set of internal nets. Experimental cylindrical submerged cages have been used in the Caribbean, and other mechanical means of raising and lowering various cage designs have been tried in Canada and Spain (Fish Farming International, 1990).

Artificial Reefs Artificial reefs can provide a good environment for both finfish and shellfish. Reefs have been designed and used to enhance natural production, to provide a focal point for stock enhancement, and to provide an attachment point for aquaculture operations. Reefs that have been designed for purposes not directly related to fisheries also attract fish. Other primary uses of reefs include control of beach erosion; provision of recreational diving sites; and disposal of municipal wastes, scrap, and solid waste. Reef systems, therefore, can be classified either as structures constructed specifically for the purpose of fisheries enhancement or as structures built for other purposes that also attract fish. Reefs constructed for fisheries enhancement efforts are usually located in public waters. Because they are publicly owned waters, assigning harvest rights is difficult; as a result, public reefs offer little to the aquaculture entrepreneur in the United States.

Artificial reefs are being explored outside the United States both as a medium for increasing commercial fisheries and for their potential use in aquaculture activities (Fabi et al., 1989; Grove et al., 1989). Fabi et al. (1989) demonstrated potential for rope culture of bivalve shellfish in association with artificial reefs in areas outside those normally used for shellfish culture. On a larger scale, Grove et al. (1989) describe a "marine ranch" in Japan consisting of artificial reefs, feeding and support stations, and onshore hatchery and nursery rearing of fish that are then released into the area after they have been trained or conditioned to respond to acoustic signals for feeding.

Use of Deep Ocean Water One futuristic system that could operate either onshore or offshore involves the coupling of marine aquaculture operations with ocean thermal energy conversion (OTEC) projects that are currently being engineered and tested (ECOR, 1989). Cold nutrient-rich deep ocean water is pumped up from below an OTEC facility for energy conversion and then passed through an onshore culture system, allowing the growth of various plant species, finfish, crustaceans, or mollusks. In addition, the use of cold water from the deep ocean offers the potential for cultivation of cold water species in tropical climates.

Thermal power plants sited in the marine environment often use seawater for thermal cooling, thus producing and discharging large quantities of heated water. Use of this water for marine aquaculture is being practiced in Europe and the United States for certain site-specific applications. This arrangement has limitations, however. The water may be too warm, or the "plumbing" may not be suitable. Moreover, power plants are subject to shutdowns, which cause fluctuations in temperature—a condition that generally precludes aquaculture.

A system in operation in Hawaii does not use the cold water for energy conversion, but instead provides the water directly for cultivation of cold

Microalgae production facility (Cyanotec) located on the OTEC facility in Kona, Hawaii.

water species. The system was funded by the state and is used for development of new technologies and culture systems. This aquaculture park, as it is called, has umbrella permits for effluents, land use, and other requirements, and thus allows for a streamlined approach to efforts to develop new species, technologies, and culture systems. Another system designed for use in Japan similarly utilizes the nutrient-rich deep ocean waters (ECOR, 1989).

Harvest and Postharvest Technology

Advances in harvest and postharvest technology can have significant positive impact on the commercial success of marine aquaculture through associated benefits such as:

- increased shelf life of fish products;
- diversification or increase in range of marketable products;
- control and monitoring of the quality and safety of products; and
- reduction in labor costs.

Technological approaches can help marine aquaculture to provide high-quality, safe products, as well as enable more efficient, economical, and adaptable marketing strategies. Most of the needs discussed below require the modification or improvement of an existing technology or the transfer of a technology successfully used in another industry. It should be noted that most aspects of preprocessing, processing, packaging, and transport to market apply not only to marine aquaculture but to freshwater aquaculture and wild-caught fish as well.

Harvest

Innovative harvest technology can contribute to improved product quality, increased yield, and reduced labor costs. In some situations, environmental benefits also can be realized, such as the elimination of escapement.

The development of methods to distinguish male from female salmon in order to harvest each at the most appropriate time is an example of using technology to increase yield. The male salmon ideally would be harvested earlier than the female, but no technology exists for distinguishing between them nondestructively, quickly, and economically. Such technology would be advantageous for other species as well.

Product quality can be enhanced by the development of rapid and accurate quality testing methods (e.g., disease identification and detection of

Power reel and seine used to harvest fish from earthen ponds.

heavy metals or human pathogens). In Alaska, development of the shellfish industry has been hampered by the lack of such quality testing technology for naturally occurring toxins. Once developed, the technology would serve other shellfish producers as well.

The cost of shrimp harvest could be reduced by improved seining (harvesting by net) equipment that would decrease labor requirements and potentially provide for size selection as well. Similarly, improved technology to harvest shellfish mechanically could reduce labor costs and also reduce environmental impacts. Mechanization of the harvest of finfish can benefit producers also. In addition to reducing cost, technology can help minimize the stress to fish during harvest and thereby maintain the quality of the product.

Preprocessing

A fundamental objective of aquaculture is to provide nutritious, high-quality seafood. To meet this objective, postharvest technologies and methodologies are needed that will ensure the preservation of quality. Higher product quality could be realized if economical means that are available for rapid precooling and for cooling and holding other food products were applied to the fish and shellfish industry. For example, during warm weather, up to two days can be required to properly cool oysters in a truck. This slow cooling results in significant quality loss. Better methods exist for cooling and holding or bagging oysters, but the cost of these systems typically is too high for commercial application.

A significant loss of product is common when finfish are eviscerated. For example, the recovery rate for gutted salmon with the head is 87-88 percent, and without the head 71 percent, leaving potential by-products that could be used as new products such as fish or livestock feed. The opportunity already exists for efficient and economical use of by-products to minimize waste and increase product yield.

The shelf life of fish products could be improved by surface sterilization methods (e.g., disinfectant dip or irradiation) that reduce microbial load on the surface of seafood. The surfaces of fish that have been cut, such as product made into fillets, or of peeled shrimp, can have a large microbial load due to exposure and handling. Some foreign companies allow low-dosage irradiation in order to kill microorganisms on the surface. The use of irradiation for this application has not been approved in the United States, but it is approved for other products such as spices and potatoes (to prevent sprouting).

Current technologies for peeling and deveining shrimp are detrimental to both the quality and the yield of shrimp. Improved mechanical methods of shucking shellfish and of peeling and deveining shrimp can reduce labor costs and possibly increase yield. The mechanical peeling method uses a great deal of water. Consequently, quality is reduced as soluble proteins and flavor are leached from the flesh; yield (meat) is lost; and significant levels of water effluent are discharged to the environment. Mechanical methods for shucking shellfish are used, but only on those that are to be cooked. Other methods have been used experimentally for shucking (e.g., high-pressure steam), but they are not very effective. Methods also are needed for sizing and grading to ensure uniform fish tailored for specific markets (e.g., white tablecloth restaurants want specific sizes and shapes).

Processing and Packaging

The primary objectives of processing and packaging are the preservation of quality and the extension of shelf life. Many traditional methods of preservation could be improved through technical advancement. Salt, sugar, and pH could be lowered, for example. Shelf life can be increased by controlling the gas content in the package using modified atmosphere and controlled atmosphere packaging (MAP/CAP). In addition to ensuring longer shelf life, technology can contribute to increased yield and improved market appeal; product losses or deterioration due to dehydration and dripping can be minimized with the right combination of materials, gases, and temperature control.

Packaging and labeling of fish as an aquaculture product can enable consumers to identify the product. Aquaculture producers have a greater ability to control the environment and inputs for their product than do wild fish producers, so they have the capacity to ensure consistency in product quality and safety through differentiation in packaging and labeling. The opportunity also exists for development of new products, new product forms (i.e., salmon sold as pan-size fish), and value-added products (e.g., smoked fish).

Transport to Market

Proper transport of fish and fish products is critical to achieving high-quality product. More precise temperature and time controls are needed to maintain quality and improve shelf life of seafood. To maximize the shelf life, refrigeration at low temperatures is necessary because temperature rise can cause significant loss of quality. Refrigerated trucks and temperature monitoring and control devices are available but are not used consistently in many segments of the aquaculture industry because of their costs. Enforced standards for transport may encourage better use of this technology.

Many consumers in the United States prefer to buy live fish and crustaceans and are willing to pay a premium if the fish arrive at the market in good condition. Technologies for feeding, aeration, and maintaining water quality would facilitate live transport to domestic and international markets. The Chinese have developed methods of transport that allow them to ship live fish to Japan by boat, taking three weeks, with a survival rate comparable to an overnight air shipment from the United States. Japanese researchers have also developed technology to ship live animals to Japan from other countries: live shrimp are shipped in wet sawdust, and acupunctured fish are reported to be shipped live from New Zealand and Australia (Thompson, 1992). Live transport by ship could help the U.S. producer compete with foreign producers.

OTHER RESEARCH AND ENGINEERING OPPORTUNITIES

Marine Fisheries Enhancement

Advances in a number of technological areas would benefit both public stock enhancement and private ocean ranching. For example, improved hatchery techniques for production of red drum, striped bass, oysters, scallops, and other species would contribute to the success of marine fisheries enhancement of endangered and threatened species for recreational and commercial fisheries in the Gulf of Mexico, the central and south Atlantic, San Francisco Bay, and along the Atlantic and Gulf coasts. A number of other species could benefit from the development of new hatchery production techniques. Candidate species include haddock, halibut, codfish, red snapper, flounder, snook, and tarpon.

Water reuse systems have particular significance for private ocean ranching of salmon, where the production of smolts requires large quantities of fresh water of appropriate quality and temperature. Such water supplies are limited and generally already committed to other uses. Where adequate water supplies are available, they often are not well located for effective management of the return and harvest of adults. An ideal release site would be one where no other salmon runs exist nearby and where harvest in the ocean can be managed in an appropriate manner (Bevin, 1988). An effective water reuse technology system would make it possible to rear and release salmon with small, isolated water supplies or, possibly, on the margin of the normal range of salmon streams.

Improved understanding is necessary to establish the biological design criteria and the biological behavior of species for which ocean ranching and stock enhancement are viable. Improvements in hatchery design and development must be based on an understanding of reproduction physiology and juvenile nutrition that is presently lacking for many candidate species. Research is needed, too, on the straying of released juveniles and returning adults. The relationship between hatchery and release strategies and the rate of straying is not well understood. Furthermore, the impact of straying on the genetic diversity of wild stocks has not been adequately quantified.

An improved system of marking hatchery fish that are released for enhancement and ocean ranching (akin to branding cattle) also would benefit marine fisheries management. Although technology is available to mark (or tag) small fish released into the ocean, current techniques are labor intensive and costly. In addition, the procedure for recognizing and reading the mark (i.e., coded wire or nose tag) when the adult fish are harvested, is costly and subject to uncertainty. Most current marking systems, whether tags attached to the shell of a mollusk or a wire implant for a fish, are used on only a small proportion or sample of the animals released. The labor requirements for obtaining and processing the data from returns result in a long time lag after harvesting before the information is available. Consequently, this information can be used only to evaluate what happened last week or last year and does not provide data in "real time" to assist in distinguishing between natural and hatchery-produced stocks for the purposes of fishery management. Passive integrated transponder tags have the potential to be miniaturized and produced at a reasonable cost.

In situations where fish from public hatcheries, private ocean ranching, and natural spawns are being harvested in the same waters (e.g., coastal Oregon), harvest managers often have to decide whether to manage for the natural fishery (i.e., allow very low harvest) or to allow a higher harvest and risk unacceptable losses without knowing what fish are being caught. Because the time frame in which the harvest takes place is often 24 hours or less, a manager cannot postpone decisions until small, hard-to-extract wire tags are located and read. Ideally, a mark should be easy to impose, easy to recognize on the harvested fish, easy to read without expensive special equipment or major effort, and it should not diminish the chances of the individual fish's survival (Sedgwick, 1982).

Cost allocation and recovery are also of special concern to those engaged in private marine fisheries enhancement. These concerns raise one of the most fundamental issues of ocean harvest—who owns the fish (Keen, 1988). A number of schemes have been suggested, including the idea of charging private "enhancers" for the use of ocean ranges while charging fishermen a tax on any privately produced fish or shellfish harvested. In addition, some suggest a need for compensation based on the straying of certain finfish such as salmon. All fishery stocks that are released can be expected to have some straying of individuals from their acclimatization sites, whether they have been produced naturally or in a hatchery (Mayo, 1989). Quantification of the degree of straying would be useful in defining damages, but sufficient and reliable data are not available at this time. The availability of a system for providing visible marks, as defined above, could alleviate some of the concerns surrounding harvest management, cost allocation, and straying.

Stock enhancement in a variety of forms appears to have been helpful in sustaining and rebuilding populations of salmonids, striped bass, red drum, oysters, clams, scallops, and perhaps other species in the marine environment. Nevertheless, the fact is that not all enhancement efforts are successful; historically, most have not been. Lack of success generally has been due either to poorly conceived and executed programs that target species whose biology makes them unsuitable candidates or to the lack of fully developed culture technology.

Broodstock Domestication

The utilization of wild broodstock is necessary at the initiation of a marine aquaculture operation. Consequently, continued access to public stocks of fish for brood animals is essential at the present time for several species. The use of wild broodstock to support commercial operations will probably not be acceptable for the long term because of the shortage of such stock and opposition from sports or commercial fishermen. Advantages from breeding domesticated stocks and making genetic improvements include increased disease resistance and accelerated growth, maturation, and reproduction.

The availability and predictability of access to wild broodstock are serious concerns that should decrease in importance as domesticated stocks are established. Marine aquaculture cannot develop into a truly commercial industry until it is free of dependence on wild stocks for its supply of brood animals. This has occurred to date for very few, if any, marine animals (Pacific oysters, hard clams, and Atlantic salmon, to some degree), and true domestication may take decades or more to achieve. The domestication of agricultural crops and animals has taken place over hundreds of years; attempts to domesticate marine aquatic organisms have been under way for only about 20 years at most.

In California, producers of striped bass and white sturgeon depend on access to wild fish for spawning. Although the California broodstock collection program was intended to last only 5 years, it now appears that 10 years or more may be necesary before the industry can rely on its own spawning stock (California Department of Fish and Game, 1989). Producers are anxious to develop domestic broodstock to end their dependency on public agencies for permits and to be able to implement genetic improvements. Recent spawning success with some domestically reared sturgeon gives reason for optimism.

The process of establishing founder populations for broodstock and initiating the domestication of a species requires careful attention to the fundamentals of breeding and a long-term commitment of people, facilities, and funding (Doyle, 1983; Gjerdem, 1983; Kinghorn, 1983; Lester, 1983; Refstie, 1990). Systems have to be developed for the long-term husbandry, selection, and special requirements of each broodstock, with precaution to include escape-proof features to allay concerns over genetic impacts of escapees on wild stocks.

Broodstock domestication for the future is likely to include a wide range of species. Finfish species for which broodstock domestication is imperative include striped bass and its hybrids, Pacific salmon, sturgeon, red drum, dolphin, snapper, grouper, and flounder for food fish, as well as ornamentals. The shellfish species include penaeid shrimp, clams, and oysters (particularly for disease resistance).

Biotechnology and Genetic Engineering

Production of Improved Strains

The United States has been the leader in the development of transgenic species (species carrying introduced genes) for culture purposes. Transgenic organisms may possess a variety of potential advantages including increased growth rates, disease resistance, decreased aggression, sterile progeny, increased tolerance of temperature, or other environmental conditions, and improved market characteristics. According to Kapuscinski and Hallerman (1990a), a total of 14 species of transgenic fish had been produced as of July 1989.

Other countries are already making use of the (largely American) techniques of transgenic production. In the United States, advances in this area await the establishment and implementation of a regulatory system that provides for the use of transgenic organisms in aquaculture (Hallerman and Kapuscinski, 1990; Kapuscinski and Hallerman, 1990b).

Kapuscinski and Hallerman (1990b) point out that the introduction of nonnative genes into fish is likely to affect nontarget traits as well and that the phenotypic performance of transgenic fish is virtually unknown. This is in large part due to regulatory constraints on the release of transgenic fish into outdoor production systems for cultural trials. Further, Tiedje et al. (1989) note that uncontrolled introduction of transgenic fish into natural aquatic communities should not be allowed because their ecological impacts are entirely unknown. Thus, Kapuscinski and Hallerman (1990b) recommend that the American Fisheries Society take the following positions with regard to transgenic fish:

1. Support research in such areas as "phenotypic characterization of transgenic lines, evaluation of the performance of transgenic lines, improvement

of sterilization techniques, and development of ecological risk assessment models and protocols" to provide data for rational policy decisions.

2. Advocate caution in the use of transgenic fish. No introductions of transgenic fish into production-scale aquaculture facilities should be allowed until risk assessments and demonstrations of little possibility of environmental impact have been completed. Further, "stockings of transgenic fishes into natural waters should be barred unless and until a body of research strongly indicates the merits of and ensures the ecological safety of stocking a particular transgenic fish into a particular receiving natural system."

3. Advocate regulations improving the comprehensiveness of the Coordinated Framework (National Institutes of Health (NIH) and U.S. Department of Agriculture (USDA) guidelines) in the United States. This recommendation would require that all production of transgenic species to take place under NIH guidelines and would establish mandatory federal regulatory review and authority over proposed releases and transport of transgenic fish.

The application of selected or directed breeding to aquatic organisms has been reviewed by a variety of authors (e.g., Doyle, 1983; Gjerdem, 1983; Lannan and Kapuscinski, 1986; Shultz, 1986; Gall, 1990). Breeding programs in aquaculture are generally in their infancy, but efforts have been initiated in a number of groups, especially finfish: Atlantic salmon (Friars et al., 1990; Refstie, 1990); and coho salmon (Hershberger et al., 1990). Selective breeding of mollusks has been limited principally to bivalves (Purdom, 1987; Wada, 1987).

Hybridization and polyploidy may produce culture-adapted strains. Hybridization has been documented among salmonids (for review, see Chevassus, 1979, 1983) and several groups of algae (Sanbonsuga and Neuschal, 1977, 1978; Cain, 1979; Guiry, 1984). Another biotechnical option with potential for aquaculture is the production of monosex populations (Purdom, 1983; Yamazaki, 1983; Billard, 1987). Gynogenetic or androgenetic (all female or all male) offspring can be produced, resulting in nonreproducing populations. A drawback is that the population may lack vigor due to inbreeding (Thorgaard, 1986).

Clearly, the United States has sufficient capabilties to make substantial progress in the areas of biotechnology and genetic engineering for aquaculture. Discussions among leading researchers suggest that application and adaptation of such technologies in the culture of marine species could be expected to result in numerous advances in marine aquaculture, including the following:

- accelerated growth and maturation of brood animals via ploidy manipulation, gene insertion, hormonal treatment, or other methods;

- improved culture characteristics (growth, food conversion efficiency, body composition, disease resistance, fecundity, hardiness, etc.) via selective breeding, hybridization, ploidy or sex manipulations, or transgenic techniques;
- production of 100 percent sterile organisms for commercial grow-out on farms or in pens, while reproductively competent organisms serve as brood stock;
- use of mitochondrial DNA methods or other analyses to detect low-level genetic change in cultured stocks to permit assessments of potential impacts of released hatchery animals on wild populations; and
- insertion of genes coding for particularly desirable traits (e.g., homing in salmon) into other species of cultured or "sea-ranched" animals.

Hedgecock and Malecha (1990) conclude that "it is very unlikely that genetic engineering by direct genomic intervention and modification will contribute to shrimp and prawn aquaculture in the next decade," due to the lack of basic knowledge of genes that affect production characteristics and of methods for inserting these genes into crustaceans. Biotechnology is more likely to be employed as a tool in more traditional programs, especially for establishing genetic markers, manipulating gametes via cryopreservation and chromosome number, and controlling sex.

Disease Assessment and Treatment

Disease Diagnosis

The development of diagnostic tests has been identified as one of the principal means of improving aquaculture productivity (Ratafia and Purinton, 1989). Rapid, accurate, and inexpensive techniques for disease assessment and certification for marine organisms in culture are essential prerequisites to screening large numbers of fry, fingerlings, postlarvae, or spat rapidly for certain critical diseases. Biotechnical methods, when applied to marine aquaculture, should allow the establishment of meaningful, effective state and national disease certification programs, which are critical for advancement of the industry.

Some types of diagnostic tools have improved markedly in recent years, and further advances are likely. Isoelectric focusing (Shaklee and Keenan, 1986) and mitochondrial DNA analyses (Brown and Wolfinbarger, 1989; Palva et al., 1989; Reeb and Avise, 1990) permit detailed analyses of fish and shellfish stocks and detection of minute genetic differences, sometimes even within a limited geographic area. Fatty acid composition analysis provides another way by which wild aquatic organisms can be differentiated from cultured individuals of the same species. This effectively eliminates the possibility of poaching protected wild stocks for sale as aquaculture products.

Therapeutics

Disease treatment represents another as yet underdeveloped research area. Currently, only six chemotherapeutics are approved for aquaculture use by the Environmental Protection Agency (EPA) and the FDA. Seven other chemicals, either EPA approved for aquaculture uses and exempt from FDA registration or exempt from EPA registration entirely, are also being used as chemotherapeutics (for review, see Williams and Lightner, 1988).

The most comon approach to the administration of antibiotics and other therapeutic agents is immersion (osmotic), injection, or oral intake with feed (DeCrew, 1972; Strasdine and McBride, 1979; Austin et al., 1981; Marking et al., 1988). Few vaccines have been developed for aquaculture use. Immunization against *Vibrio* spp. and related bacteria genera has been practiced among finfish culturists for several years, with treatment by immersion or intraperitoneal injection (Cipriano et al., 1983; Schiewe et al., 1988). Similar methods of immunization are now being explored for shrimp (Itami et al., 1989).

Experimental work also has been conducted on the use of gelatin capsule implants for some antibiotics (Strasdine and McBride, 1979) and on various types of water treatment, including ozonation (Wedemeyer et al., 1978; Tipping, 1987), ultraviolet irradiation, and chlorination (Bedell, 1971; Sanders et al., 1972). Formal actions by the U.S. Fish and Wildlife Service and the USDA have ensured that federal and state animal scientists, the pharmaceutical industry, and the USDA collaborate on determining needs and developing research protocols for aquaculture-related drugs, which fall under the category of minor-use animal drugs (Schmick, 1988).

As the above discussion suggests, the need and potential for improvements in disease treatment are substantial. The development of medications and immunizations is badly needed, as are improved delivery systems for antibiotics that will not result in the release of antibiotics to the rearing waters. The cost of obtaining FDA approval is a major barrier, however.

SUMMARY

Advances in technology and an improved understanding of the biology of relevant species are essential for marine aquaculture to overcome many of the major constraints on future development. Some new and improved technologies would solve specific technical problems directly and thereby improve economic feasibility; other technologies would alleviate environmental concerns and diminish conflicts with other coastal zone activities.

Many of the technical constraints on marine aquaculture can be reduced or eliminated by developing new and improved onshore and nearshore systems; developing new and improved auxiliary systems; and establishing the biological, ecological, and engineering knowledge base required for making sound decisions.

NOTE

1. The term cages and pens have been defined by Beveridge, 1987 as follows: cages are enclosed on the bottom as well as the sides, typically by mesh or net screens, whereas the bottom of pens is formed by the seabed.

REFERENCES

Aase, H. 1985. Effect of the use of flow developers in fish-rearing cages for salmon, Fisherdirektoratets, Havforskningsinstitutt, Bergen, Norway.

Alabaster, John S. 1982. Report of the EIFAC Workshop on Fish-Farm Effluents. EIFAC Technical Paper No. 41. European Inland Fisheries Advisory Commission, FAO. 186 pp.

Arnold, C.R., B. Reid, and B. Brawner. 1990. High density recirculating grow out systems. Pp. 182-184 in Red Drum Aquaculture. University of Texas, Austin, TAMU-SG-90-603.

Austin, B., D.A. Morgan, and D.J. Alderman. 1981. Comparison of antimicrobial agents for control of vibriosis in marine fish. Aquaculture 26:1-12.

Balchen, J.G. 1987. Bridging the gap between aquaculture and the information sciences. In Automation and Data Processing in Aquaculture, J.G. Balchen and A. Tysso, eds. IFAC Proceedings 1987. No. 9. Pergamon Press.

Bedell, G.W. 1971. Eradicating *Cerotomyxa shasta* from infected water by chlorination and ultraviolet irradiation. Prog. Fish-Culturist 33:51-54.

Beveridge, M.C.M. 1987. Cage Aquaculture. Farnham, England: Fishing News Books, Ltd. 332 pp.

Bevin, D. 1988. Problems of managing mixed-stock salmon fisheries. In Salmon Production, Management, and Allocation—Biological, Economic and Policy Issues, William J. McNeil, ed. Oregon State University Press.

Biedenback, J.M., L.L. Smith, T.K. Thomsen, and A.L. Lawrence. 1989. Use of the nematode *Panagrellus redivivus* as an *Artemia* replacement in a larval penaeid diet. Journal of the World Aquaculture Society 20(2):61-71.

Billard, R. 1987. The control of fish reproduction in aquaculture. Pp. 309-305 in Realism in Aquaculture: Achievements, Constraints, Perspectives, M. Bilio, H. Rosenthal, and C. Sindermann, eds. Breden, Belgium: European Aquaculture Society.

Bromley, P.J., and P.A. Sykes. 1985. Weaning diets for turbot (*Scophthalmus maximus* L.), sole *(Solea solea)* and cod *(Gadus morhua* L.) Pp. 191-211 in Nutrition and Feeding in Fish, C.B. Cowery, A.M. Mackie, and J.G. Bell, eds. New York: Academic Press.

Brown, B.L., and L. Wolfinbarger. 1989. Mitochondrial restriction enzyme screening and phylogenetic relatedness in the hard shell clam genus *Mercenaria*. Part 2. Population variation. Technical Report Virginia Department of Environmental Science No. TR-89-1.

Brune, D.E. 1990. Reducing the environmental impact of shrimp pond discharge. ASAE Paper No. 90-7036. American Society of Agricultural Engineers, St. Joseph, Mich.

Brune, D.E., and R.H. Piedrahita. 1983. Operation of a retained biomass nitrification system for treating aquaculture water for reuse. Proceedings of the First International Conference on Fixed-Film Biological Processes. 845-869.

Cain, J.R. 1979. Survival and mating behavior of progeny and germination of zygotes from inter- and intraspecific crosses of *Chlamydomonas eugametos* and *C. moewusii* (chlorophycease, Volvocales). Phycologia 18(1):24-29.

California Department of Fish and Game. 1989. Private striped bass broodstock collection and rearing program: 1989 activities and eight-year progress report. Unpublished manuscript, 10 pp.

Carey, J., and B. Kramer. 1966. Fish Hatchery Design Memorandum No. 14.1. Dworshak Dam and Reservoir. Prepared for the U.S. Army Engineering District, Walla Walla, Wash.

Chevassus, B. 1979. Hybridization in salmonids: Results and perspectives. Aquaculture 17:113-128.

Chevassus, B. 1983. Hybridization in fish. Aquaculture 33:245-262.

Cho, C.Y., H.S. Bayley, and S.J. Slinger. 1974. Partial replacement of herring meal with soybean meal and other changes in a diet for rainbow trout *(Salmo gairdneri)*. J. Fish. Res. Bd. Can. 31:1523-1528.

Cipriano, R.C., J.K. Morrison, and C.E. Starliper. 1983. Immunization of salmonids against *Aeromonas salmonicida*. J. World Maricul. Soc. 14:201-211.

Clarke, R., and M. Beveridge. 1989. Offshore fish farming. INFOFISH International (3/89):12-15.

Conklin, D.E., L.R. D'Abramo, and K. Norman-Boudreau. 1983. Lobster nutrition. Pp. 413-423 in Handbook of Mariculture, Vol. I: Crustacean Aquaculture, J.P. McVey, ed. Boca Raton, Fla.: CRC Press.

Conrad, K.M., M.G. Mast, and J.H. MacNeil. 1988. Performance, yield, and body composition of fingerling channel catfish fed a dried waste egg product. Prog. Fish-Culturist 50:219-224.

Cowery, C. B., J.A. Pope, J.W. Adron, and A. Blair. 1971. Studies on the nutrition of marine flatfish. Growth of the plaice *Pleuronectes platessa* on diets containing proteins derived from plants and other sources. Marine Biology 10(2):145-153.

Davis, E.M., G.L. Rumsey, and J.G. Nickum. 1976. Egg-processing wastes as a replacement protein source in salmonid diets. Prog. Fish-Culturist 38:20-22.

DeCrew, M.G. 1972. Antibiotic toxicity, efficacy, and teratogenicity in adult spring chinook salmon (*Oncorhynchus tshawytscha*). J. Fish. Res. Bd. Can. 29(11):1513-1517.

Donaldson, E.M., U.H.M. Fagerlund, D.A. Higgs, and J.R. McBride. 1978. Hormonal enhancement growth. Pp. 456-578 in Fish Physiology, Vol. VIII, W.S. Haar, D.J. Randall, and J.R. Brett, eds. New York: Academic Press.

Doyle, R.W. 1983. An approach to the quantitative analysis of domestication selection in aquaculture. Aquaculture 33:167-185.

Electric Power Research Institute (EPRI). 1990. A summary description of the second workshop on the role of macroalgal oceanic farming in global change. July 23-24, Newport Beach, California. Electric Power Research Institute, Palo Alto.

Engineering Committee on Oceanic Resources (ECOR). 1989. Ocean Energy Systems. Report of ECOR International Working Group. Japan Marine Science and Technology Association, Tokyo.

Fabi, G., L. Fiorentini, and S. Giannini. 1989. Experimental shellfish culture on an artificial reef in the Adriatic Sea. Bulletin of Marine Science 44(2):923-933.

Flynn, G. 1990. Presentation to the committee. Halifax, Nova Scotia, June 11-15.

Forster, J. 1990. Presentation to the committee. Davis, Calif., March 19-20.

Friars, G.W., J.K. Bailey, and K.A. Coombs. 1990. Correlated responses to selection for grilse length in Atlantic salmon. Aquaculture 85:171-176.

Fridley, R.B., R.H. Piedrahita, and T.M. Losordo. 1988. Challenges in aquacultural engineering. Agricultural Engineering (May/June):12-15.

Gall, G.A.E. 1990. Basis for evaluating breeding plans. Aquaculture 85:125-142.

Gjerdem, T. 1983. Genetic variation in quantitative traits and selection breeding in fish and shellfish. Aquaculture 33:51-72.

Gowen, R.J. 1988. Release strategies for coho and chinook salmon released into Coos Bay, Oregon. In Salmon Production, Management, and Allocation—Biological, Economic and Policy Issues, William J. McNeil, ed. Oregon State University Press.

Grant, B.F., P.A. Seib, M. Liao, and K.E. Corpron. 1989. Polyphosphorylated L-ascorbic acid: A stable form of vitamin C for aquaculture feeds. Journal of the World Aquaculture Society 20(3):143-157.

Grove, R.S., C.J. Sonu, and M. Nakumura. 1989. Recent Japanese trends in fishing reef design and planning. Bulletin of Marine Science 44(2):984-996.

Guiry, M.D. 1984. Structure, life history, and hybridization of Atlantic *Gigartina teedii* (Rhodophyta) in culture. Br. Phycol. J. 19(1):37-55.

Hallerman, E.M., and A.R. Kapucinski. 1990. Transgenic fish and public policy: Regulatory concerns. Fisheries (Bethesda) 15 (1):12-20.

Halver, J.E. 1988. Fish Nutrition. New York: Academic Press. 489 pp.

Hedgecock, D., and S.R. Malecha. 1990. Prospects for the application of biotechnology to the development of shrimp and prawns. In Shrimp Culture in North America and the Caribbean, P.A. Sandifer, ed. Advances in World Aquaculture, World Aquaculture Society, Baton Rouge, La.

Hershberger, W.K., J.M. Myers, R.N. Iwamoto, W.C. McAuley, and A.M. Saxton. 1990. Genetic changes in the growth of coho salmon *(Oncorhynchus kisutch)* in marine net-pens produced by ten years of selection. Aquaculture 85:187-198.

Holt, G.J. 1992. Experimental studies of feeding of larval red drum. Journal of the World Aquaculture Society. (In press.)

Holt, G.J. 1990. Growth and development of red drum eggs and larvae. Pp. 46-50 in Red Drum Aquaculture. TAMU-SG-90-603. Texas A&M Sea Grant College Program, College Station.

Huguenin, J.E., and J. Colt. 1989. Design and Operating Guide for Aquaculture Seawater Systems. Developments in Aquaculture and Fisheries Science, 20. New York: Elsevier Publishing. 264 pp.

Imai, T. (ed.) 1977. Aquaculture in Shallow Seas: Progress in Shallow Sea Culture (translated from Japanese). National Oceanic and Atmospheric Administration, Washington, D.C. 615 pp.

Institution of Civil Engineers (ICE). 1990. Proceedings of the Conference on Engineering for Offshore Fish Farming, Glasgow, Scotland, October 17-18. London: Thomas Telford.

Itami, T., Y. Takahashi, and Y. Nakamura. 1989. Efficacy of vaccination against vibriosis in cultured Kuruma prawns *Penaeus japonicus*. J. Aquatic Animal Health 1:238-242.

Kaiser, G.E., and F.W. Wheaton. 1991. Engineering aspects of water quality monitoring and control. Pp. 210-232 in Engineering Aspects of Intensive Aquaculture. Proceedings from the Aquaculture Symposium, Cornell University, Ithaca, N.Y. Ithaca: Northeast Regional Agricultural Engineering Service.

Kanazawa, A., S. Koshio, and S. Teshina. 1989. Growth and survival of larval red sea bream, *Pagrus major* and Japanese flounder, *Paralichthys olivaceus* fed microbound diets. Journal of the World Aquaculture Society 20(2):31-37.

Kapuscinski, A.R., and E.M. Hallerman. 1990a. Transgenic fish and public policy: Anticipating environmental impacts of transgenic fish. Fisheries (Bethesda) 15(1):2-11.

Kapuscinski, A.R., and E.M. Hallerman. 1990b. Transgenic fishes: AFS Position Statement. Fisheries (Bethesda) 15(4):2.5.

Keen, E. 1988. Ownership and Productivity of Marine Fishery Resources. Blacksburg, Va.: The McDonald and Woodward Publishing Company.

Kerr, N.M., M.J. Gillespie, S.T. Hull, and S.J. Kingwell. 1980. The design, construction, and location of marine floating cages. Pp. 70-83 in Proceedings of the Institute of Fisheries Management Cage Fish Rearing Symposium, University of Reading. London: Janssen Services.

Ketola, H.G. 1975. Requirement of Atlantic salmon for dietary phosphorus. Trans. Amer. Fish. Soc. 104(3):543-551.

Ketola, H.G. 1982. Effect of phosphorus in trout diets on water pollution. Salmonid 6(2):12-15.

Ketola, H.G. 1985. Mineral nutrition: Effects of phosphorus in trout and salmon feeds on water pollution. Pp. 465-473 in Nutrition and Feeding of Fish, C.B. Cowery, A.M. Mackie, and J.G. Bell, eds. London: Academic Press.

Ketola, H.G. 1988. Salmon fed low-pollution diet thrive in Lake Michigan. U.S. Fish and Wildlife Service, Research Information Bulletin No. 62-25, April.

Ketola, H.G. 1990. Studies on diet and phosphorus discharges in hatchery effluents. Abstract, International NSMAW Symposium, Guelph, Canada, June 5-9.

Ketola, H.G., H. Westers, C. Pacor, W. Houghton, and L. Wubbels. 1985. Pollution: Lowering levels of phosphorus, experimenting with feed. Salmonid 9(2):11.

Ketola, H.G., M. Westers, W. Houghton, and C. Pecor. 1990. Effects of diet on growth and survival of coho salmon and on phosphorus discharges from a fish hatchery. American Fisheries Society Symposium No. 11.

Kinghorn, B.P. 1983. A review of quantitative genetics in fish breeding. Aquaculture 31(2,3,4):283-304.

Korringa, P. 1976. Farming Cupped Oysters of the Genus *Crassostrea*. A Multidisciplinary Treatise. Amsterdam: Elsevier. 224 pp.

Kruner, G., and H. Rosenthal. 1983. Efficiency of nitrification in trickling filters using different substrates. Aquacultural Engineering 2:49-67.

Lannan, J.E., and A.R.D. Kapuscinski. 1986. Application of a genetic fitness model to extensive aquaculture. Aquaculture 57:81-87.

Lannan, J.E., R.O. Smitherman, and G. Tchobanoglas, eds. 1985. Principles and Practices of Pond Aquaculture. Corvallis, Ore.: Oregon State University Press.

Lester, L.J. 1983. Developing a selective breeding program for penaeid shrimp mariculture. Aquaculture 33:41-50.

Linfoot, B.T., and M.S. Hall. 1987. Analysis of the motions of scale-model, sea-cage systems. In Automation and Data Processing in Aquaculture, J.G. Balchen and A. Tysso, eds. IFAC Proceedings. No. 9. New York: Pergamon Press.

Losordo, T.M. 1991. Engineering consideration in closed recirculating systems. Pp. 58-69 in Aquaculture Systems Engineering. Proceedings of the World Aquaculture Society and the American Society of Agricultural Engineers Jointly Sponsored Session at the World Aquaculture Society Meeting, San Juan, Puerto Rico.

Losordo, T.M., R.H. Piedrahita, and J.M. Ebeling. 1988. An automated data acquisition system for use in aquaculture ponds. Aquacultural Engineering 7:265-278.

Lovell, T. 1989. Nutrition and Feeding of Fish. New York: Van Nostrand Reinhold. 260 pp.

Malone, R.F., and D.G. Burden. 1988. Design of recirculating blue crab shedding systems. Louisiana Sea Grant College Program, Center for Wetland Research, Louisiana State University, Baton Rouge.

Marking, L.L., G.E. Howe, and J.R. Crowther. 1988. Toxicity of erythromycin, oxytetracycline, and tetracycline administered to lake trout in water baths by injection or by feeding. Prog. Fish-Culturist 50:197-201.

Mayo, R.D. 1989. A Review of Water Reuse. World Aquaculture Society Annual Meeting, Los Angeles.

McLean, E., and R. Ash. 1990. Modified uptake of the protein antigen horseradish peroxidase (HRP), following oral delivery of rainbow trout, *Oncorhynchus mykiss*. Aquaculture 87(3/4):373-380.

Meek, R. 1990. Presentation to the committee. Davis, Calif., March 19-20.

Miller, G.E., and G.S. Libey. 1985. Evaluation of three biological filters suitable for aquacultural applications. J. World Maricul. Soc. 16:158-168.

National Research Council (NRC). 1974a. Nutrient Requirements of Trout, Salmon, and Catfish. Board on Agriculture and Renewable Resources. Washington, D.C.: National Academy Press.

National Research Council (NRC). 1974b. Research Needs in Animal Nutrition. Board on Agriculture and Renewable Resources. Washington, D.C.: National Academy Press.

National Research Council (NRC). 1977. World Food and Nutrition Study: Panel on Aquatic Food Resources. Commission on International Relations. Washington, D.C.: National Academy Press.

National Science Foundation (NSF). 1991. Workshop on Engineering Research Needs for Off-Shore Mariculture Systems. East-West Center, University of Hawaii, September 26-28.

Nettleton, J.A. 1990. Comparing nutrients in wild and farmed fish. Aquaculture Magazine 16(1):34-41.

Neuschal, M. 1990. Presentation to the committee. Davis, Calif., March 19-20.

Palmer, J.E. (ed.) 1989. The application of artificial intelligence and knowledge-based systems techniques to fisheries and aquaculture workshop report. Virginia Sea Grant Publication 89-03.

Palva, T.K., H. Lehvaeslaiho, and E.T. Palva. 1989. Identification of anadromous and non-anadromous salmon stocks in Finland by mitochondrial DNA analysis. Aquaculture 81(3/4):237-244.

Piedrahita, R.H. 1991. Modeling water quality in aquaculture ecosystems. In Aquaculture and Water Quality, D.E. Brune and J.R. Tomass, eds. Baton Rouge, La.: World Aquaculture Society.

Purdom, C.E. 1983. Genetic engineering by the manipulation of chromosomes. Aquaculture 33:287-300.

Purdom, C.E. 1987. Methodology on selection and intraspecific hybridization in shellfish—A critical review. Pp. 285-282 in Selection, Hybridization, and Genetic Engineering in Aquaculture. Vol. 1, K. Tiews, ed. Heenemann Verlagsgellschaft mbh, Berlin.

Ratafia, M., and T. Purinton. 1989. Emerging aquaculture markets. Aquaculture Magazine (July/August):32-46.

Reeb, C.A., and J.C. Avise. 1990. A genetic discontinuity in a continuously distributed species: Mitochondrial DNA in the American oyster, *Crassostrea virginica*. Genetics 124(2):397-406.

Reed, B. 1989. Evaluation of a recirculating raceway system for the intensive culture of the penaeid shrimp *Penaeus vannamei* Boone. M.S. thesis, Department of Biology, Corpus Christi State University, Texas.

Refstie, T. 1990. Application of breeding schemes. Aquaculture 85:163-169.

Richards, G.P. 1988. Microbial purification of shellfish: A review of depuration and relaying. Journal of Food Protection 51 (3):218-251.

Rogers, G.L., and S.L. Klemetson. 1985. Ammonia removal in selected aquaculture water reuse biofilters. Aquacultural Engineering 4:135-154.

Sanbonsuga, Y., and M. Neuschal. 1977. Cultivation and hybridization of giant kelps (Phaeophyceae). Pp. 91-96 in Proccedings of the Ninth International Seaweed Symposium, A. Jensen and J. Stein, eds. Princeton: Science Press.

Sanbonsuga, Y., and M. Neuschal. 1978. Hybridization of *Macrocystis* (Phaeophyta) with other float bearing kelps. J. Phycol. 14(2):214-224.

Sanders, J.E., J.L. Fryer, D.A. Leith, and K.D. Moore. 1972. Control of the infectious protozoan *Ceratomyka shasta* by treating hatchery water supplies. Prog. Fish-Culturist 34(1):13-17.

Santulli, A., E. Puccia, and V. D'Amelio. 1990. Preliminary study on the effect of short-term carnitine treatment on nucleic acids and protein metabolism in sea bass (*Dicentrarchus labrax* L.) fry. Aquaculture 87(1):85-90.

Schiewe, M.H. A.J. Novotny, and L.W. Harrell. 1988. Vibriosis of salmonids. Pp. 323-327 in Disease Diagnosis and Control in North American Marine Aquaculture, C.J. Sindermann and D.V. Lightner, eds. Amsterdam: Elsevier.

Schmick, R.A. 1988. The impetus to register new therapeutants for aquaculture. Prog. Fish-Culturist 50:190-196.

Sedgwick, S. 1982. The Salmon Handbook. London: Andre Deutsch, Ltd.

Shaklee, J.B., and C.P. Keenan. 1986. A practical laboratory guide to the techniques and methodology of electrophoresis and its application to fish fillet identification. CSIRO Marine Laboratories, Report 177. 59 pp.

Shigueno, K., and S. Itoh. 1988. Use of Mg-L-ascorbyl-2-phosphate as a vitamin C source in shrimp diets. Journal of the World Aquaculture Society 19(4):168-174.

Shultz, F.T. 1986. Developing a commercial breeding program. Aquaculture 57: 65-76.

Strasdine, G.A., and J.R. McBride. 1979. Serum antibiotic levels in adult sockeye salmon as a function of route of administration. J. Fish. Biol. 15:135-140.

Thompson, T. 1992. Boxed fish tank can deliver live U.S. seafood to Japan. Journal of Commerce, January 3, p. 6B.

Thorgaard, G.H. 1986. Ploidy manipulation and performance. Aquaculture 57:57-64.

Tiedje, J.M., R.K. Colwell, Y.L. Grossman, R.E. Hodson, R.E. Lenski, R.N. Mack, and P.J. Regal. 1989. The planned introduction of genetically engineered organisms: Ecological considerations and recommendations. Ecology 70:298-315.

Tipping, J. 1987. Use of ozone to control ceratomyxosis in steelhead trout *(Salmo gairdneri).* Ozone Science and Engineering 9:149-152.

Wada, K.T. 1987. Selective breeding and intraspecific hybridization—molluscs. Pp. 293-302 in Selection, Hybridization, and Genetic Engineering in Aquaculture, Vol. 1, K. Tiews, ed. Berlin: Keenemann Verlagsgellschaft mbh.

Wang, J.K. 1988. Shared resources aquatic production systems. American Society of Agricultural Engineers, ASAE Paper No. 88-5001. St. Joseph, Mich.

Wang, J.K. 1990. Managing shrimp pond water to reduce discharge problems. Aquacultural Engineering 9:61-73.

Weaver, D. 1990. Presentation to the committee. Halifax, Nova Scotia, June 11-15.

Wedemeyer, G.A., N.C. Nelson, and C.A. Smith. 1978. Survival of salmonid viruses infectious hematopoietic necrosis (IHNV) and infectious pancreatic necrosis (IPNV) in ozonated, chlorinated, and untreated waters. J. Fish. Res. Bd. Can. 35:875-879.

Weston, D.P. 1986. The Environmental Effects of Floating Mariculture in Puget Sound. School of Oceanography, College of Ocean and Fishery Sci., University of Washington, Seattle. 148 pp.

Williams, R.R., and D.V. Lightner. 1988. Regulatory status of therapeutics for penaeid shrimp culture in the United States. Journal of the World Aquaculture Society 19(4):188-196.

Wyban, J.A., and E. Antill, eds. 1989. Instrumentation in aquaculture. Proceedings of a Special Session at the World Aquaculture Society 1989 Annual Meeting, Oceanic Institute, Hawaii. 101 pp.

Yamada, S., Y. Tanaka, M. Sameshima, and Y. Ito. 1989. Pigmentation of prawn *(Penaeus japonicus)* with carotenoids. Effect of dietary astaxanthin, β-carotene, and canthaxanthin on pigmentation. Aquaculture 87(3/4):323-330.

Yamazaki, F. 1983. Sex control and manipulation in fish. Aquaculture 33:329-354.

6

Information Exchange, Technology Transfer, and Education

OVERVIEW

The development model for an agricultural product typically follows a logical and time-proven sequence. Information is developed, tested in field conditions, monitored by agencies, extended to users, and then transferred to the private sector to be incorporated into commercial activities. Investors and lending institutions ultimately rely on a support system to make the logical sequence yield benefits. The Morrill Act of 1862 provides the foundation of the U.S. agricultural support system through the establishment of the land grant university system of higher education. A wide spectrum of technical disciplines is represented at each land grant campus. The agricultural support system includes an integrated research, teaching, and extension effort involving faculty researchers, extension personnel, and the private sector.

Research and teaching in each state occur at campus locations and numerous experimental stations. Research results are extended to agricultural producers and the agricultural business infrastructure by a network of Cooperative Extension Service (CES) centers that includes on-campus specialists, as well as staff located in every county. CES technical training techniques include short courses and workshops, which together with communication on an individual basis, have proved to be effective means of research interpretation and technology transfer. Publications and on-site demonstrations at cooperating grower farms add to the process of continuing education. Extension programs provide technical assistance on systems design, production, business planning, management, and marketing.

Aerial view of the James M. Waddell, Jr., Marine Research and Development Center in Bluffton, South Carolina.

Support of aquaculture is provided by universities involved with land grant or sea grant programs. Unlike agriculture, however, the land grant system's linkage of campus research, experimental farms, and organization for information and technology transfer has not been duplicated in the area of marine aquaculture. Specifically appropriated funds for the National Sea Grant College Program partially address extension, but establishment of experimental "farms" and extension positions specifically in marine aquaculture are limited and hamper the effort to transfer the agricultural model to the aquatic area. At the present time, the processes for information exchange, technology transfer, and education in the marine aquaculture field are to a large degree ad hoc and informal. A more structured system is needed to facilitate the flow of information and the transfer of new technology and to provide more formal training. Information may be communicated verbally or in written form as raw data, printed material, or personal consultation. The important point is that information is of use to a field of endeavor only if it is shared.

The transfer of technology is a process that requires more extended effort than information exchange, usually involving animals or facilities. Technology transfer is generally accomplished through a practitioner who is a specialist and frequently occurs through private consultations, short courses, demonstration or pilot projects, and the systems of higher education. Private consultants working with clients are constantly pushing at the edge of technology to make operations more cost-effective and profitable. However, consultants often are limited in the amount of technology they can transfer from project to project by the proprietary nature of their work.

Short courses conducted by university, extension, and private companies are frequently an effective means of technology transfer. Demonstration projects, such as those at the Waddell Mariculture Center, the Oceanic Institute, the Texas A&M and University of Texas Marine Science Institute, Louisiana State University, and the University of Washington, provide valuable technology transfer points when the potential aquaculture entrepreneur is aware of their existence.

In addition to information exchange and technology transfer to practitioners, two areas of education that need attention in relation to marine aquaculture are (1) increasing public awareness of aquaculture and (2) training of specialists. Very little, if any, general knowledge of marine biology or aquaculture is transmitted in kindergarten through twelfth grade. In fact, awareness of marine systems in general, at those levels, is rarely stressed. At the undergraduate level, current training for a career in aquaculture is primarily through the selection of technical electives; rarely, through an option in animal science or biology; or through two- and four-year technical programs. Graduates from these programs, however, often lack sufficient background in more multidisciplinary training that includes engineering, economics, property/environmental law, and other subjects related to the marine environment that would enable them to work independently without additional training. The majority of graduate programs include specialties such as fisheries or aquaculture within another discipline (e.g., marine biology/science, animal science, or engineering). In these programs, as at the undergraduate level, it is difficult to obtain specific training in marine aquaculture.

Continuing education can provide both training for specialists in areas other than those involved in their usual work activities and disseminate information about new techniques or technologies to practitioners. Private and public short courses or continuing education courses, workshops and seminars, annual trade association and professional meetings, and educational exchange programs are the primary techniques used.

To improve the present situation of information exchange, technology transfer, and education, fundamental prerequisites for a formal system of expert information and technology need to be established. The prerequisites include a structured system for the exchange of basic information and the transfer of technology, formal educational programs, support of continuing education and training for specialists, and public education. The creation of these structures and mechanisms would improve the practice of marine aquaculture and would benefit the economics and marketing aspects of industry operations as well. A structured system would stimulate more rapid advances in all aspects of marine aquaculture systems through better dissemination of research results.

INFORMATION EXCHANGE

Currently, information is communicated through a number of processes and institutions with varying degrees of success. The vast majority of information transfer within the marine aquaculture community occurs through trade journals, trade shows, professional journals, scientific meetings, and extension programs. Private consultants are involved in information transfer, but they are limited, as a rule, by the client-professional relationship. Consultants are participants in trade show programs however and, because of their technical training, at professional society meetings as well. As consultants, they frequently present general information, comment on technology adaptability, and convey general trends as opposed to disclosing specific project information.

Sea Grant Program extension personnel also conduct information exchange activities in marine aquaculture, but they are responsible for the entire spectrum of marine activities. This limits the commitment of time and the relative continuity with which they can address marine aquaculture concerns. Sea Grant extension personnel are assigned geographically (by county or region) and focus on local information transfer and education opportunities. Their technical support system includes campus-based specialists in the fields of aquaculture, biology, engineering, law, economics, and others. Specialists, although well educated, frequently do not have the option of working exclusively on a single aspect of marine resource use. Many are specialists in a technical discipline and have limited training or experience specific to marine aquaculture.

The Regional Aquaculture Centers established in 1985 under the auspices of the U.S. Department of Agriculture (USDA) are responsible for both the generation and the dissemination of information. Research projects are funded at public research institutions, and results are exchanged via Cooperative Extension (USDA) and Sea Grant Marine Advisory Service (National Oceanic and Atmospheric Administration) (NOAA) extension programs. The communication of research results and of information and technology needs between scientists and commercial practitioners is a vital role of these extension services. The opportunity exists for increased involvement of marine aquaculture agents/advisors and specialists from the USDA Regional Aquaculture Centers to adequately provide for the feedback aspect of effective information exchange, as well as to provide the information delivery system.

The National Aquaculture Information Center (NAIC), affiliated with the

USDA National Agriculture Library, has the potential to serve as the clearinghouse for information and to become a focal point for information transfer. However, the withdrawal of base-level funding support for the NAIC has severely limited its capability to retrieve and disseminate information. The lack of a funded central information retrieval system affects the ability of professionals responsible for information exchange to do the best possible job. Extension personnel with limited marine aquaculture training would benefit from access to such a system. Availability of an international aquaculture reference and data network would be particularly beneficial.

Current methods of data collection and statistics from national and international sources are inadequate in terms of both quantity and quality. The only national data source for aquaculture, *Aquaculture Situation and Outlook Reports* by USDA, covers freshwater species fairly accurately but does not include many of the cultured marine species. Data for the production of new and evolving species are often not reported. In addition, a significant proportion of cultured production may be marketed in such a way that it is not reported (USDA, 1988).

The International Fisheries Office of the National Marine Fisheries Service (NMFS) occasionally reports statistics and developments in marine aquaculture for selected countries. These irregularly produced country analyses are useful for near-term outlook purposes. They are an information source of increasing value as the United States strives to meet international competition. Information transfer would be more complete if a larger number of countries were reviewed each year.

Two recent publications attempt to present data on aquaculture production and value for all freshwater and marine aquaculture species. The first, *The Potentials of Aquaculture: An Overview and Bibliography* (Hanfman et al., 1989), put out by the National Aquaculture Information Center, states that accurate data on current U.S. production of salmon are unavailable. In addition, statistics on the status of worldwide aquaculture are based on data from the United Nations Food and Agriculture Organization (FAO). As of early 1991, the most current statistics used in FAO publications were from 1987 or, in some cases, part of 1988. The dramatic expansion of marine aquaculture production in the past three years renders these statistics inadequate for assessing the status of current production. The most accurate current world production statistics must be culled from trade journals, an approach that is time consuming and, in some cases, incomplete.

A second source of aquaculture statistics and one that offers the most complete data set on the western region of the United States was compiled by the Western Regional Aquaculture Consortium (Chew and Toba, 1991) in cooperation with the USDA. The authors offer caveats on their statistics as follows: "The report is . . . a compilation of estimates in aquaculture production as provided by . . . [industry] representatives." They then explain that the statistics represent averages of several estimates and that the probability exists for errors. The most complete information available on aquaculture production and value for the western region is, it seems, preliminary at best.

A number of actions could improve the overall task of information exchange. The two exchange components are dissemination of research results and data distribution. The dissemination component is hindered by the relatively small number of individuals with an assigned marine aquaculture extension responsibility. The agents need opportunities to receive professional improvement training and to have access to accurate information retrieval systems. Specialists, similarly, need these opportunities to be better prepared, with particular emphasis on international developments and formal interaction with researchers.

The second component of the information exchange process, data distribution, begins with the understanding that research and trade developments yield data. A national network for electronic collection and transfer of data, such as that employed for the national crop reporting system, would vastly improve data collection. In addition, the designation of a state specialist to provide leadership and be a point of contact for information, followed later by county agents when the industry develops (i.e., to lead, organize, and follow with numbers), would be effective in gathering state-level information. This system could be patterned after the successful process used for agricultural data collection and transfer (USDA, 1990).

Specific mechanisms can be applied to attain a system that will facilitate the free flow of information and the transfer of technology generated nationwide, specifically in the areas identified above. Increased and improved (i.e., accurate) data collection to enhance the status of marine aquaculture is basic to all aspects of marine aquaculture development (Hanfman et al., 1989). Existing data collection mechanisms (i.e., the USDA crop and fisheries reporting services) could provide industry data on production, employment, gear, economics, and trade, for example. The transfer of information—data, unpublished reports, and literature—from foreign countries to the United States could be accomplished through use of the NAIC as well as through U.S. scientists' visits abroad.

TECHNOLOGY TRANSFER

Because the success of marine aquaculture depends on technology development, both nationally and internationally, frequent needs arise to acquire and/or transfer new technology rapidly to fulfill investment plans and remain competitive. These economic pressures create incentives to transfer technology prematurely, which often results in outcomes that are initially unsatisfactory. Modifications "on-site" to improve the technology depend on extended relationships and contacts between those transferring and those receiving the technology. Such situations require continuing relationships among extension personnel, researchers, and the individuals, companies, or agencies that apply their findings.

The implications of this pattern for the Sea Grant Marine Advisory Service and Cooperative Extension are that arrangements will be required for prolonged contact among the parties. This situation is in contrast to the information exchange function that deals with large numbers of users for brief periods (i.e., meetings, short courses). An improved technology transfer system for marine aquaculture will require an organizational structure and resources that allow extended relationships with users and, consequently, will require the allocation of more resources to this process than at present.

The adoption of technology by industry users often results in modification or adaptations of the technology. Thus, the process of technology transfer may lead to technology modifications that are proprietary. As a result, a close, confidential relationship is needed among all parties. Recognition of this potential development is necessary prior to technology transfer agreements. University offices of technology transfer have experience with the proprietary aspects of improvements. Specific attention to proprietary rights is important because of the complexities and the basic role of technology transfer in a changing industry. The anticipated beneficial feedback of research needs to campuses when technology is transferred may not occur otherwise.

A mechanism of providing statewide specialists for facilitating information and technology transfer could be expanded upon and offered within some concentration of the Sea Grant Program and the Cooperative Extension Service. Even though there are some pilot-scale demonstration projects around the United States, substantial benefits can be obtained from developing marine aquaculture field stations and demonstration of semicommercial-scale production. Semicommercial production demonstration facilities would bridge the present gap between the small-scale, purely applied research technology or pilot-scale production and the commercial production technology.

Currently, the USDA Regional Aquaculture Centers are working to improve the exchange and integration of the results of collaborative applied research among university, extension, and industry researchers and users. These efforts have met with varying degrees of success in various centers. New methods of integration would serve to increase the rate of development of marine aquaculture in the United States by minimizing unnecessary duplication of efforts.

A large number of innovations in marine aquaculture are developed by other countries. Current state-of-the-art technology around the world could be exchanged and transferred to enable faster growth of the world marine aquaculture industry. The means for international technology transfer and exchange of scientists is not well organized. A program to encourage U.S. scientists' visits abroad could help to bridge the gap in international technology transfer. Because marine aquaculture production operates in a global market, care must be taken to emphasize the benefits to the world marine aquaculture industry. New initiatives could also be taken to develop additional foreign exchange programs.

EDUCATION

The availability of formal education specific to marine aquaculture is limited, so the field draws practitioners from the general areas of science and engineering. Additional recruiting efforts may be necessary to increase the number of graduate students entering these programs who intend to use their skills in the United States if a strong base of talent is to be available for the development of marine aquaculture.

In addition to higher enrollments, some form of cross-training or interdisciplinary studies needs to be incorporated in the training and education programs of aquaculturists. Aquaculture is a profession that requires expertise in a broad number of subject areas, such as biology, engineering, economics, regulation, administration, business, trade, and marketing. In designing cross-disciplinary programs, however, care must be taken to ensure that the end result is not a half-trained engineer or a half-trained biologist. The goal is to train an engineer, for example, who has some rudimentary knowledge of the relevant biological concepts and terminology, and is capable of formulating appropriate questions to ask a biologist. The same concept would apply to the biologist, economist, or social scientist specializing in aquaculture.

A number of universities and other public organizations offer short or continuing education courses in marine aquaculture. In addition, private organizations conduct short courses in marine aquaculture-related topics. The Cooperative Extension Service and the Sea Grant Marine Advisory Service conduct workshops and seminars for private aquaculturists and agency personnel. Annual trade shows and professional meetings also provide educational opportunities in specific areas. The ongoing United States-Japan cooperative education exchange program is effective in providing technology transfer and exchange between these two countries, and could serve as a model for exchanges with other countries.

Public information on and awareness of aquaculture are limited and, in some cases, consist primarily of information about widely publicized negative environmental effects. A reasonable goal for public education efforts is to raise the level of awareness about aquaculture of the general public, people involved in the industry, and regulators and agency personnel.

At the kindergarten through twelfth grade level, educational activities could be focused on marine systems in general. Audiences could be addressed through 4-H, designating an Ocean Week, and similar programs. At the undergraduate level, increased educational opportunities in biological systems engineering would serve to round out one of the important areas of training that is currently inadequate. Expanded graduate programs focused on areas of marine aquaculture in engineering, veterinary medicine, economics, and policy/regulatory programs would provide professionals who could apply their expertise toward marine aquaculture. In addition, foreign exchange programs for graduate students and research scientists would both serve as training and educational programs and increase the opportunities for technology transfer and exchange.

Overall awareness of the industry itself, some of its unique problems, and the need for collaborative efforts among diverse professions can be improved through the promotion of linkages among professional organizations, government agencies, and interest groups, such as the NMFS, the U.S. Fish and Wildlife Service (FWS), recreational and commercial fishermen, environmental groups, consumer groups, agricultural sectors, the American Society of Agricultural Engineers, the American Fisheries Society, and the World Aquaculture Society. A broader group of people needs to be reached with information regarding the marine aquaculture industry in order to raise the awareness of others and ensure information exchange.

MARINE AQUACULTURE TECHNOLOGY CENTERS

Marine aquaculture would benefit from the establishment of marine aquaculture technology centers to serve as facilities where multidisciplinary teams would come together to carry out focused development, advancement, and technology transfer of marine aquaculture systems. Such centers would complement the biological research laboratories in which the basic biological understanding and design criteria must originate. Such centers would not require new facilities at new sites, but the reprogramming and redevelopment of existing facilities and research institutions or the use of existing USDA, NMFS, or FWS facilities.

Candidate facilities that could be transformed into regional marine aquaculture technology centers include state and federal centers that can be found in all regions of the country that have an interest in coastal and marine matters, so that geographic and species considerations could be easily encompassed. These centers could be designated without any additional funding, although consideration might be given to allocation of an appropriate portion of the Saltonstall-Kennedy funds (raised from duties on food imports, including aquaculture products) to support research that would reduce U.S. dependency on imported seafood. Institution of a peer review process similar to the National Science Foundation's competition for basic research grants would ensure a high level of quality for research programs conducted at the centers.

The following characteristics would be appropriate for any national centers for marine aquaculture technology:

- centers at three or more locations to provide for a range of climatic and ecological conditions (e.g., Atlantic, Pacific, Gulf);
- programs focused on building on existing technology in a timely manner;
- facilities for a wide range of research areas (e.g., biological investigations, marine structures, marine materials, sensing and monitoring equipment, water reuse technologies, processing methods, quality enhancement; and tagging and marking systems). Biological laboratories should complement (not duplicate) existing research institutions;
- a strong staff of professionals from various disciplines including biologists, engineers, economists, food scientists, institutional/policy specialists, and marketing/business specialists;
- an advisory board that includes strong representation from commercial aquaculture;
- direct electronic information exchange and interaction with other research programs at universities and state or federal agencies; and
- modern facilities designed for maximum flexibility in responding to new research directions, including capability to undertake remote research with portable equipment.

Such a program would contribute to the advancement of marine aquaculture with appropriate attention to economic, institutional, and environmental concerns by assembling an interdisciplinary staff, directing their activities into areas of specific interest to commercial aquaculturists, and providing facilities for research and extension activities. Advancement would be stimulated by the interdisciplinary, targeted research and the associated technology transfer.

The technology centers could be complemented by research parks that would provide the private sector with a location for conducting research and development for commercialization, Scrivani (1990) suggested the concept of state aquaculture parks where entrepreneurs could lease space, seawater, and infrastructure and be covered by an umbrella permit. Such parks would foster commercial operations, but even more importantly, would foster commercialization (i.e., parks could play an important role in technology transfer). A planned linkage between the technology centers and such aquaculture parks would facilitate the deployment of new technology.

SUMMARY

Current information and technology transfer efforts for marine aquaculture are ad hoc and informal, compared to those for traditional agriculture. They lack dedicated resources and structure. The leading federal agencies with responsibilities for aquaculture extension services have few aquaculture specialists and especially few marine aquaculture specialists. Yet the technology in this field is evolving rapidly, and researchers and practitioners are, of necessity, involved in informal national and international data and technology transfer. A more coordinated, formalized, and carefully structured system would overcome some of the problems of duplication of effort and unreliable or unavailable data, and would provide a solid base for the development of advanced technologies necessary to the success of this industry.

The enhancement of cooperative extension programs to provide training and the establishment of academic programs aimed specifically at the needs of marine aquaculture are also prerequisites to its advancement. Enhancement of national marine aquaculture in the United States could be assisted by emphasis on increased collaboration between the Cooperative Extension Service and the Sea Grant Marine Advisory Service. An expansion of the NAIC within the National Agriculture Library—to include more information on marine aquaculture, using the most modern techniques and real-time data available, and announcing its availability—would increase the awareness of regulators, managers, lenders, and investors, and their ability to serve the industry. As a result, investment and regulation would occur in the best possible circumstances. Such a system also would provide researchers and research program administrators with the best available information on which to base research proposals.

REFERENCES

Chew, K.K., and D. Toba. 1991. Western region aquaculture industry: Situation and outlook report. Western Regional Aquaculture Consortium, University of Washington, Seattle. 23 pp.

Hanfman, D.T., S. Tibbitt, and C. Watts. 1989. The potentials of aquaculture: An overview and bibliography. U.S. Department of Agriculture, Bibliographies and Literature of Agriculture No. 90. Washington, D.C.

Scrivani, P. 1990. Presentation to the committee. Davis, Calif., March 19.

U.S. Department of Agriculture (USDA). 1988. Aquaculture Situation and Outlook Report. Economic Research Service. AQUA 1. Washington, D.C.

U.S. Department of Agriculture (USDA). 1990. Outlook for U.S. Agricultural Exports. Foreign Agriculture Service, Economic Research Service, Washington, D.C.

7

Conclusions and Recommendations

CONCLUSIONS

Marine aquaculture—including the farming of marine finfish, shellfish, crustaceans, and seaweed, as well as ocean ranching of anadromous fish—is a rapidly growing industry in many parts of the world. In the United States, freshwater aquaculture (primarily the farming of catfish, trout, and crayfish) is an expanding industry; however, marine aquaculture has yet to sustain more than limited economic success. Based on its investigations, the committee concluded that a number of benefits would accrue to the nation from a healthy marine aquaculture industry, including wholesome food to replace harvests of wild fish from stocks that are declining or at maximum sustainable yield, products for export to improve the nation's balance of trade, enhancement of commercial and recreational fisheries and of fisheries that are utilized fully, economic opportunities for rural communities, and new jobs for skilled workers. Advancement of the science and technology base in marine aquaculture also provides potential benefits to other industries, such as biotechnology and pharmaceuticals.

Constraints on the industry have included difficulties and costs of using coastal and ocean space; public concerns about environmental effects of wastes on water quality; conflicts with other users of the coastal zone (e.g., boaters and fishermen); increasing population with concomitant increases in pressure on coastal areas; a limited number of sites with suitable water quality; objections to marine aquaculture installations on aesthetic grounds from coastal property owners; broad ecological issues involving concerns about genetic dilution of wild stocks and transfer of diseases through the escape of cultured animals; and a limited understanding of the biological criteria needed for the design of viable systems.

On the other hand, the consumption of seafood in the United States is increasing at the same time that yields from capture fishing are reaching the limits of sustainable yield and the nation relies increasingly on imports to meet the growing consumer demand for seafood. The opportunity, therefore, exists for U.S. aquaculture to develop the capability to supply this growing demand and for marine aquaculture to make a significant contribution.

Although legislation to promote aquaculture was passed in 1980 and again in 1985 (National Aquaculture Act, P.L. 96-362) (National Aquaculture Improvement Act, P.L. 99-198), a number of problems have prevented these expressions of policy intent from effectively transforming marine aquaculture into a dynamic industry. First, except for the establishment of the regional aquaculture centers, no funds were ever appropriated to agencies to implement the provisions of these acts. Second, the needs of marine aquaculture have tended to be overshadowed by the interests of the freshwater aquaculture industry, which are more closely linked to those of the traditional agriculture community through its geographic focus in inland farming areas. Moreover, marine aquaculture, because of its location in the coastal zone, faces the more complex ocean regulatory regime, as well as widespread public interest and concern about activities that take place in or near the ocean. U.S. marine aquaculture is unlikely to reach its full potential until substantial changes are made in the ways in which federal and state governments support and regulate these activities and until environmental concerns are addressed.

The present study investigated the opportunities for improving the outlook for U.S. marine aquaculture and concluded that the issues that constrain development will need to be specifically addressed through three primary avenues: (1) advances in the scientific, technical, and engineering base that underlies this industry, both to achieve more cost-effective operations and to mitigate environmental problems; (2) changes in federal and state agency roles to provide a regulatory and funding framework that encourages the industry's growth while ensuring that environmental concerns are addressed; and (3) congressional action to address a number of unresolved policy issues and to clearly define a national policy. For marine aquaculture to succeed in the United States, a more active and forceful federal role will be needed, one that employs a wider range of incentives for aquaculture development and that centralizes authority (and corresponding resources) to support the promotional role of the lead federal agency, the U.S. Department of Agriculture (USDA).

Achieving the objectives outlined above and discussed in detail in the following recommendations will depend on active oversight of the executive agencies that presently are charged with implementing the national policies expressed in the National Aquaculture Act and the National Aquaculture Improvement Act. Such oversight, to be effective, must come from

Congress, through a committee or subcommittee with responsibility for ensuring that executive agencies coordinate their aquaculture-related activities to achieve the maximum efficiency in the use of limited resources and that sufficient funds are appropriated to carry out the legislative mandate.

RECOMMENDATIONS

Based on the committee's conclusions, the following recommendations are made with the aim of fostering the emerging marine aquaculture industry and enabling it to establish a sound base from which to move forward in the future.

Advances in Technology and Engineering— A Marine Aquaculture Initiative

The opportunity exists for technology and increased knowledge to provide solutions to many of the environmental, economic, and biological limitations that constrain marine aquaculture's transformation into a significant U.S. industry. However, the opportunity can be realized only if federal policy and action strongly support the development of new technology and the research necessary to provide the biological information for technology design.

The committee recommends that Congress make a $12 million national commitment to a strategic R&D initiative to develop marine aquaculture technology and the biological understanding needed to address environmental issues and concerns, and to provide economical systems. Leadership in this initiative should be provided by the U.S. Department of Agriculture, with coordination by the Joint Subcommittee on Aquaculture (JSA) and implemented under memoranda of understanding (MOUs) among federal agencies involved in marine aquaculture regulation and research. The initiative should include research and development to address the following:

- the interdisciplinary development of environmentally sensitive, sustainable systems that will enable significant commercialization of onshore (on land) and nearshore marine aquaculture without unduly increasing conflict over use of the coastal area;
- development of the knowledge base for technologies and candidate species needed to make decisions regarding commercialization of offshore marine aquaculture operations that avoid the environmental impacts of nearshore operations;
- creation of (1) technology centers to be used for the above technology development programs and (2) marine aquaculture parks with umbrella permits for marine aquaculture for fostering development and deployment of new environmentally sensitive commercial technology;
- design and implementation of improved higher-education programs and

procedures, and systems to collect and exchange data and technical information; and

- promotion of marine aquaculture as a vital component of fisheries stock enhancement by (1) facilitating aquaculture's role in the preservation of threatened or endangered species populations and of genetic diversity, including the involvement of private sector facilities; (2) developing production procedures for the broader range of species necessary for effective mitigation of negative impacts on fish and shellfish stocks; and (3) developing and implementing improved methods for determining the effectiveness of using cultured stock for fish and shellfish enhancement activities in support of commercial, recreational, and ecological purposes.

Federal Agency Responsibilities and Actions

The federal agencies with primary jurisdiction over marine aquaculture activities include the U.S. Department of Agriculture (USDA), the Fish and Wildlife Service (FWS), and two branches of the National Oceanic and Atmospheric Administration (NOAA)—the National Marine Fisheries Service (NMFS) and the National Sea Grant College Program. USDA was designated as lead agency in the National Aquaculture Act Amendments of 1985. The Joint Subcommittee on Aquaculture provides a forum for these and other federal agencies to discuss their aquaculture activities. It is appealing to envision a highly efficient centralization of all responsibility for marine aquaculture in one agency—USDA. However, consideration of the realities of longstanding and traditional jurisdictional responsibilities of other agencies—FWS for hatcheries, NOAA for activities that take place in the oceans, the National Science Foundation (NSF) for funding basic research—undermine the feasibility of such an approach. On the other hand, more active leadership and more effective coordination of federal activities are necessary to translate the intent of existing national legislation regarding aquaculture into greater commercialization. The following recommendations are aimed at achieving action on behalf of marine aquaculture, and will require executive direction and congressional oversight to ensure that they are implemented.

U.S. Department of Agriculture

It is recommended that the lead role of USDA be strengthened by creation under its auspices of several concise and comprehensive interagency memoranda of understanding that clarify the mission, role, and responsibilities of each agency with respect to aquaculture and specifically marine aquaculture. These MOUs should spell out the mutual understanding necessary to create an environment for collaboration on needs and issues related to marine aquaculture, and they must be reinforced with strong, high-level executive direction. With the MOUs in place, it is anticipated that the JSA can make a substantial contribution in influencing the actions of the participating agencies, going far beyond the role of coordination and information exchange now achievable.

It is recommended that the USDA be charged with leadership in the promotion of commercial aquaculture including the research and support services (i.e., National Aquaculture Information Center) required, particularly in the areas of production, processing, distribution, and marketing of marine aquaculture products, especially as food products. The leadership role should involve:

- promotion of marine aquaculture as a provider of wholesome food;
- support of related R&D programs and establishment of related facilities for bringing new and improved systems and new species to commercial feasibility;
- reinstatement of base-level support for the National Aquaculture Information Center (NAIC); and
- collection and dissemination of production and marketing statistics and related information.

To provide effective leadership for marine aquaculture, USDA will need to establish a formal entity focused on aquaculture, specifically including marine aquaculture at an appropriately high level within the agency, and to acquire expertise in marine aquaculture throughout its offices. Specific additional funds should be allocated for targeting marine aquaculture activities in existing USDA programs such as the Agricultural Research Service, the Agricultural Extension Service, and the Economic Research Service.

Joint Subcommittee on Aquaculture

It is recommended that in addition to its current role as a forum for interagency discussion, the JSA be charged with designing a streamlined planning and permitting process for marine aquaculture activities emphasizing joint local, state, and federal coordination, and take responsibility for promoting the inclusion of marine aquaculture in the Coastal Zone Management Act.

Major unresolved issues that prevent marine aquaculture from achieving success should be addressed within JSA by the following actions:

- Formulate a plan for the explicit inclusion of marine aquaculture interests and impacts in coastal and offshore planning activities and in policies of state and federal agencies.

• Design a model for local, state, and federal intergovernmental review of marine aquaculture projects to coordinate the permitting process, as well as state and federal regulation of marine aquaculture. Such a model will include guidelines for (1) planning the development of marine aquaculture business parks through provision of federal incentives; (2) assessing the potential impacts of marine aquaculture development on the marine environment and anticipating conflicts with competing uses of public resources and waters; and (3) developing appropriate mitigation measures for unavoidable impacts.

• Conduct a comprehensive evaluation of impacts of the Lacey Act (P.L. 97-79, as amended in 1981) on marine aquaculture; formulate national or regional policies, guidelines, and procedures for importation and use (including release) of nonindigenous and genetically altered species and for health-related evaluation and certification; and make recommendations to Congress for appropriate changes in the Lacey Act to specifically encourage development of marine aquaculture based on ecologically sound considerations.

U.S. Fish and Wildlife Service

It is recommended that the FWS continue to exercise leadership in the area of fisheries enhancement of anadromous species. Such leadership should include:

• promoting the use of private aquaculture for enhancement of stocks of various anadromous species that are heavily fished or otherwise threatened or endangered;

• supporting the development of technology for rearing and releasing anadromous stocks where needed; and

• administering the introduction and transfer of nonindigenous anadromous species.

National Oceanic and Atmospheric Administration/ National Marine Fisheries Service

It is recommended that NOAA/NMFS be charged with leadership in the management and assessment of stock-enhanced marine fisheries. Such leadership should include:

• evaluating the effectiveness of existing and future stock enhancement programs;

• supporting the development of technology for (1) producing juvenile stocks needed for nonanadromous marine fisheries enhancement and related aquaculture, and (2) releasing marine stocks, where needed;

- assessing the impact (or potential impact) of various nearshore and offshore marine aquaculture practices on the marine environment and fisheries; and
- administering the introduction and transfer of nonindigenous marine species.

National Oceanic and Atmospheric Administration/ National Sea Grant College Program

The Sea Grant Program has supported research relevant to marine aquaculture; however, a major initiative should be undertaken in the context of environmental issues, the basic biology of candidate species, and competing uses of resources. It is recommended that NOAA/Sea Grant be charged with leadership in support of research and extension programs on marine aquaculture-related topics focused on preservation of the marine environment, understanding the life history of candidate species, and multiple use of marine resources, including associated social, economic, and policy issues. Candidate research topics include:

- environmentally safe technology, methods, and systems for culturing marine species in the marine environment;
- marine aquaculture technology that is synergistic with other uses of the sea (i.e., multiple use technologies);
- life history and developmental biology of candidate species;
- the socioeconomic dynamics of the marine aquaculture industry (e.g., effects on local employment patterns);
- methods for addressing and resolving conflicts between marine aquaculture and other competing users of the marine environment;
- comparative studies of state practices regarding the regulation and promotion of marine aquaculture; and
- alternative institutional and policy structures for managing marine aquaculture in other countries.

Congressional Action

The development of marine aquaculture is beset with complexity that stems from unique factors that distinguish it from other kinds of agricultural activity. These are:

- its interaction with other marine and coastal activities and interests—interactions often characterized by conflict;
- the fact that although marine aquaculture is ocean based, it is dependent on the use of land and freshwater resources as well; and

• the numerous environmental and regulatory considerations that are involved in the development and use of coastal zone land and water.

This complexity entails the involvement of a number of federal, state, and local agencies that are responsible for all aspects of the advocacy, promotion, conduct, and regulation of marine aquaculture, leading to an array of planning acts, policies, and regulations. For marine aquaculture to realize its potential, it must be addressed explicitly within a coordinated and coherent policy framework in federal, regional, and state ocean and coastal zone planning activities, to ensure its proper consideration and evaluation with respect to both resource development objectives and environmental impacts. Designation of marine aquaculture as a recognized coastal use under the Federal Coastal Zone Management Act and inclusion in state coastal management plans would be the first steps toward recognizing its role as a positive marine economic activity and streamlining the regulatory requirements that must be complied with to engage in a marine aquaculture enterprise.

Although most of the recommendations outlined above can be implemented by the designated agencies through MOUs and by the JSA under existing legislation (the National Aquaculture Act of 1980 and the National Aquaculture Improvement Act of 1985), three unresolved policy issues need to be addressed through new legislation and, therefore, require congressional action.

Completion of Federal Policy Framework for Marine Aquaculture

Coastal Zone To realize its potential, marine aquaculture must be explicitly included in coastal zone plans that ensure its proper consideration and evaluation in development and environmental decisions. It is recommended that Congress designate marine aquaculture as a recognized use of the coastal zone in the Coastal Zone Management Act (P.L. 92-583, as amended in 1990, P.L. 101-508). Such designation will stimulate states to include marine aquaculture in their coastal management plans for achieving a balanced approach to land use, resource development, and environmental regulation.

Federal Waters Currently, no formal framework exists to govern the leasing and development of private commercial activities in public waters under federal jurisdiction. A predictable and orderly process for ensuring a fair return to the operator and to the public for the use of public resources is necessary to the development of marine aquaculture. It is recommended that Congress create a legal framework to foster appropriate development, to anticipate potential conflicts over proposed uses, to assess potential environmental impacts of marine aquaculture, to develop appropriate mitigation measures for unavoidable impacts, and to assign fair public and private rents and returns on such operations.

Revision of Laws That Impede Development of Marine Aquaculture

The Lacey Act (P.L. 97-79, as amended in 1981) Environmental preservation and the protection of indigenous species are important concerns; however, the Lacey Act, as presently constituted, creates a barrier to the development of marine aquaculture. Control points for regulating the movement of living fish between states should be based on scientific and ecological information rather than solely on state borders. It is recommended that Congress make appropriate changes in the Lacey Act based on a comprehensive evaluation by the JSA of ecological and economic impacts to encourage the development of marine aquaculture within an environmentally sound regulatory framework.

Creation of a Congressional Committee or Subcommittee on Aquaculture

As human demand for seafood exceeds sustainable yield from traditional fisheries, dependence on capture fisheries is likely to shift to dependence on aquaculture. No mechanism currently exists for congressional policymakers to anticipate this transition and make appropriate policy decisions; nor is there a mechanism for congressional oversight of the federal agency and JSA actions mandated by the National Aquaculture Act and its amendments. It is recommended that Congress consider creating an oversight committee or subcommittee on aquaculture to provide a formal linkage between the House Agriculture Committee and the House Merchant Marine and Fisheries Committee to ensure the implementation of existing and future policies enacted to promote aquaculture.

CONCLUSION

A number of benefits will accrue to the nation from the addition of an economically vital, technologically advanced, and environmentally sensitive marine aquaculture industry. The prospects of this emerging enterprise are for healthy and vigorous growth, given a fair share of support for the development of an advanced scientific and engineering base, along with a reasonable and predictable regulatory framework. On this basis, the environmental problems that presently constrain marine aquaculture are likely to be resolved so that it can contribute to the continued vitality of the nation's living marine resources.

Bibliography

Aase, H. 1985. Effect of the use of flow developers in fish-rearing cages for salmon, Fisherdirektoratets, Havforskningsinstitutt, Bergen, Norway.

Ackefors, H., and A. Sodergren. 1985. Swedish experiences of the impact of aquaculture on the environment. International Council for the Exploration of the Sea, C.M., Vol. 40. 7 pp.

Aiken, D. 1990. Commercial aquaculture in Canada. World Aquaculture 21(2):66-75.

Aiken, D. 1990. Shrimp farming in Ecuador: An aquaculture success story. World Aquaculture 21(1): 7-10, 12-16.

Alabaster, J.S. 1982. Report of the EIFAC Workshop on Fish-Farm Effluents. EIFAC Technical Paper No. 41. European Inland Fisheries Advisory Commission, Food and Agriculture Organization. 186 pp.

Allen, K., ed. 1981. BioEngineering Symposium for Fish Culture. AFS-FCS Publication 1. 189+ pp.

Allen, S.K., P.S. Gagnon, and H. Hidu. 1982. Induced triploidy in the soft-shell clam. Journal of Heredity 73:421-428.

Anders, N.L., V.J. Norton, and I.E. Strand. An Evaluation of Potential Export Markets for Selected U.S. Fish Products. Technical Report, Maryland Sea Grant Program, Pub. No. UM-SG-TS-82-03. 89 pp.

Anderson, J., and J. Wilen. 1986. Implications of private salmon aquaculture on prices, production, and management of salmon resources. American Journal of Agricultural Economics 68(4):866-879.

Anderson, R.K., P.L. Parker, and A.C. Lawrence. 1987. A $^{13}C/^{12}C$ tracer study of the utilization of presented feed by a commercially important shrimp, *Penaeus vannamei* in pond growout system. Journal of the World Aquaculture Society 18(3):148-155.

Anonymous. Seaweed Raft and Farm Design in the United States and China. New York Sea Grant Institute.

Aoki, T., and T. Kitao. 1985. Detection of transferable R plasmids in strains of the fish-pathogenic bacterium *Pastuerella piscicida*. Journal of Fish Diseases 8: 345-350.

Aquaculture Project Group of the National Marine Fisheries Service. 1985. The Outlook for Salmon and Shrimp Aquaculture Products in the World Markets. Report prepared for the Assistant Administrator for Fisheries. 109 pp.

Armstrong, M. S., C. E. Boyd, and R. T. Lovell. 1986. Environmental factors affecting flavor of channel catfish from production ponds. Prog. Fish-Culturist 48:113-119.

Arnold, C.R., B. Reid, and B. Brawner. 1990. High density recirculating grow out systems. Pp. 182-184 in Red Drum Aquaculture. Texas A&M University, College Station.

Arnold, C.R., G.J. Holt, and P. Thomas. 1988. Red Drum Aquaculture, Proceedings of a Symposium on the Culture of Red Drum and Other Warm Water Fishes. Contributions in Marine Science, Supplement to Vol. 30, 197 pp.

Aspen Research and Information Center. 1981. A Directory of Federal Regulations Affecting Development and Operation of Commercial Aquaculture. Aspen Systems Corporation, Rockville, Md.

Aspen Research and Information Center. 1981. Aquaculture in the United States: Regulatory Constraints. Aspen Systems Corporation, Rockville, Md.

Austin, B. 1985. Antibiotic pollution from fish farms; Effects on aquatic microflora. Microbiological Sciences 2(4):113-117.

Austin, B., D.A. Morgan, and D.J. Alderman. 1981. Comparison of antimicrobial agents for control of vibriosis in marine fish. Aquaculture 26:1-12.

Avault, J.W., L.W. de la Bretonne, and J. Huner. 1975. Two major problems in culturing crawfish in ponds. Pp. 139-148 in Proceedings 2nd International Symposium on Freshwater Crawfish, J. W. Avault, ed.

Bailey, C. 1988. The social consequences of tropical shrimp mariculture development. Ocean and Shoreline Management 11:31-44.

Bailey, R. 1988. Third world fisheries: Prospects and problems. World Development 16:751-757.

Bakos, J. 1987. Selection breeding and intraspecific hybridization of warm water fishes. Pp. 303-312 in Selection, Hybridization, and Genetic Engineering in Aquaculture, vol.1, K. Tiews, ed. Berlin, Heenemann Verlagsgellschaft mbh.

Balchen, J.G. 1987. Bridging the gap between aquaculture and the information sciences. In Automation and Data Processing in Aquaculture, J.G. Balchen and A. Tysso, eds. IFAC Proceedings 1987. No. 9. Pergamon Press.

Barker, J.C., J.L. Chesness, and R.E. Smith. 1974. Pollution Aspects of Catfish Production—Review and Projections. U.S. Environmental Protection Agency Report Number EPA-660/2-74-064, National Technical Information Service PB-244-943, 121 pp.

Beal, B.F. 1988. Pp. 980-983 in Public Aquaculture in Downeast Maine: The Soft-Shell Clam Story. University of Maine at Machias.

Beam, M. 1987. Building Cages for Fish Farming. Extension Extra, South Dakota State University, U.S. Department of Agriculture. 4 pp.

Bedell, G.W. 1971. Eradicating *Cerotomyxa shasta* from infected water by chlorination and ultraviolet irradiation. Prog. Fish-Culturist 33:51-54.

Bergheim, A., and A. R. Selmer-Olsen. 1978. River pollution from a large trout farm in Norway. Aquaculture 36:267-270.

Bergheim, A., A. Silversten, and A.R. Selmer-Olsen. 1982. Estimated pollution loadings from Norwegian fish farms. I. Investigations 1978-1979. Aquaculture 28:347-361.

Bergheim, A., H. Hustveit, A. Kittlesen, and A.R. Selmer-Olsen. 1984. Estimated pollution loadings from Norwegian fish farms. I. Investigations 1978-1979. Aquaculture 28:157-168.

Bettencourt, S.U., and J.L. Anderson. 1990. Pen-Reared Salmonid Aquaculture in the Northeastern United States. U.S. Department of Agriculture, Northeast Regional Aquaculture Center Report 100. Kingston, R.I.

Beveridge, M.C.M. 1987. Cage Aquaculture. Farnham, England: Fishing News Books, Ltd. 332 pp.

Bevin, D. 1988. Problems of managing mixed-stock salmon fisheries. In Salmon Production, Management, and Allocation—Biological, Economic and Policy Issues, William J. McNeil, ed. Oregon State University Press.

Bidwell, C.A., C.L. Chrisman, and G.S. Lisbey. 1985. Polyploidy induced by heat shock in channel catfish. Aquaculture 51:25-32.

Biedenback, J.M., L.L. Smith, T.K. Thomsen, and A.L. Lawrence. 1989. Use of the nematode *Panagrellus redivivus* as an *Artemia* replacement in a larval penaeid diet. Journal of the World Aquaculture Society 20(2):61-71.

Billard, R. 1987. The control of fish reproduction in aquaculture. Pp. 309-350 in Realism in Aquaculture: Achievements, Constraints, Perspectives, M. Bilio, H. Rosenthal, and C. Sindermann, eds. European Aquaculture Society, Breden, Belgium.

Binkowski, F.P., and S.I. Doroshev. 1985. Epilogue: A perspective on sturgeon culture. Pp. 147-152 in North American Sturgeons: Biology and Aquaculture Potential, Binkowski, F.P., and S.D. Doroshev, eds., Dordrecht: Dr. W. Junk Publishers.

Boersen, G., and H. Westers. 1986. Waste solids control in hatchery raceways. Prog. Fish-Culturist 48:151-154.

Bondari, K. 1986. Response of channel catfish to multi-factor and divergent selection of economic traits. Aquaculture 57(1-4):163-170.

Bowden, G. 1981. Pp. 236-241 in Coastal Aquaculture Law and Policy: A Case Study of California. Boulder, Colo.: Westview Press.

Boyce, J. 1990. A Comparison of Demand Models for Alaska Salmon, Department of Economics, University of Alaska, Fairbanks, under contract with Fisheries Research and Enhancement Division, Alaska Department of Fish and Game. 102 pp.

Boyd, C.E. 1978. Effluents from catfish ponds during fish harvest. J. Environ. Qual. 7:59-62.

Boyd, C.E., R.P. Romaire, and E. Johnston. 1979. Water quality in channel catfish production ponds. J. Environ. Qual. 8:423-429.

Braun, L. 1988. Spirulina: Food for the Future. Aqua-Topic. U.S. Department of Agriculture, Washington, D.C. 9 pp.

Braun, L.M., and A.T. Young. 1988. Algae Culture and Uses: Microalgae. Quick Bibliography Series. U.S. Department of Agriculture, Washington, D.C. 9 pp.

Brenn, G., B. Brenig, G. Horstgen-Schwark, and E.L. Winnacker. 1988. Gene transfer in tilapia *(Oreochromis niloticus)*. Aquaculture 68:209-219.

British Columbia Marketing Branch of the Ministry of Agriculture and Food. 1986. The Market for Farmed Salmon: An Overview. ISBN 0-7726-0510-6. 21 pp.

Bromley, P.J., and P.A. Sykes. 1985. Weaning diets for turbot (*Scophthalmus maximus* L.), sole *(Solea solea)* and cod *(Gadus morhua* L.). Pp. 191-211 in Nutrition and Feeding in Fish, C.B. Cowey, A.M. Mackie and J.G. Bell, eds. New York: Academic Press.

Browdy, C.L., J.R. Richardson III., C.O. King, A.D. Stokes, J.S. Hopkins, and P.A. Sandifer. 1990. IHHN virus and intensive culture of *Penaeus vannamei*: Effects of stocking and water exchange rates on production and harvest size distribution. World Aquaculture Society, World Aquaculture 90, Abstract T17.3.

Brown, B.L., and L. Wolfinbarger. 1989. Mitochondrial Restriction Enzyme Screening and Phylogenetic Relatedness in the Hard Shell Clam Genus *Mercenaria*. Part 2. Population Variation. Technical Report, Virginia Department of Environmental Science No. TR-89-1.

Brown, J.R., R.J. Gowen, and D.S. McLusky. 1987. The effect of salmon farming on the benthos of a Scottish sea loch. J. Exp. Mar. Biol. Ecol. 109:39-51.

Brune, D.E. 1990. Reducing the Environmental Impact of Shrimp Pond Discharge. American Society of Agricultural Engineers, ASAE Paper No. 90-7036. St. Joseph, Mich.

Brune, D.E., and C.M. Drapcho. 1991. Fed pond aquaculture. Pp. 15-33 in Aquaculture Systems Engineering: Proceedings of the World Aquaculture Society and American Society of Agricultural Engineers Jointly Sponsored Session. American Society of Agricultural Engineers, St. Joseph, Mich.

Brune, D.E., and A.G. Eversole. 1989. Impact of crawfish on pond nutrient dynamics. Presented at Aquaculture '89, Los Angeles, Calif., February 12-16, 1989.

Brune, D.E., and R.H. Piedrahita. 1983. Operation of a retained biomass nitrification system for treating aquaculture water for reuse. Proceedings of the First International Conference on Fixed-Film Biological Processes. 845-869.

Burrell, V.G., Jr. 1985. Oyster culture. Pp. 235-273 in Crustacean and Mollusk Aquaculture in the United States, J.V. Huner and E.E. Brown, eds. Westport, Conn.: AVI Publishing Co.

Burrows, R. 1964. Effects of accumulated excretory products on hatchery reared salmonids. U.S. Sports Fisheries and Wildlife Research Report 66.

Burrows, R. and R. Combs. 1968. Controlled environments for salmon propagation. Prog. Fish-Culturist 30(3):57-85.

Cacho, D.J., V. Hatch, and H. Kinnucan. 1990. Bioeconomic analysis of fish growth: Effects of dietary protein and ration size. Aquaculture 88(3/4):223-238.

Cain, J.R. 1979. Survival and mating behavior of progeny and germination of zygotes from inter- and intraspecific crosses of *Chlamydomonas eugametos* and *C. moewusii* (chlorophycease, Volvocales). Phycologia 18(1):24-29.

California. 1988. A Guide to California State Permits, Licenses, Laws and Regulations Affecting California's Aquaculture Industry. Interagency Committee for Aquaculture Development, Sacramento.

California Aquaculture Association. 1990. Strategic Plan. Sacramento.

California Department of Fish and Game. 1989. Private striped bass broodstock collection and rearing program: 1989 activities and eight-year progress report. Unpublished manuscript, 10 pp.

Campbell, J.W. 1973. Nitrogen excretion. In Comparative Animal Physiology, C.L. Prosser, ed. Philadelphia: W.B. Saunders.

Canadian Aquaculture. 1990. World salmon production. f(2)2.

Carey, J., and B. Kramer. 1966. Fish Hatchery Design Memorandum No. 14.1. Dworshak Dam and Reservoir. Prepared for the U.S. Army Engineering District, Walla Walla, Wash.

Chaiton, J.A., and S.K. Allen. 1985. Early detection of triploidy in the larvae of Pacific oysters, *Crassostrea gigas*, by flow cytometry. Aquaculture 48:35-43.

Chamberlain, G.W. 1986. 1985 Growout research. Coastal Aquaculture 3(2):7-8.

Chamberlain, G.W. 1991. Status of shrimp farming in Texas. Pp. 36-57 in Shrimp Culture in North America and the Caribbean, P.A. Sandifer, ed. Baton Rouge, La.: The World Aquaculture Society.

Chamberlain, G.W., R.J. Miget, and M.G. Haby (compilers). 1990. Red Drum Aquaculture. Texas A&M University Sea Grant College Program, College Station No. TAMU-SG-90-603.

Chaurrout, D. 1980. Thermal induction of diploid gynogenesis and triploidy in the eggs of the rainbow trout (*Salmo gairdneri*, Richardson). Reprod. Nutr. Dev. 20:727-733.

Chesapeake Bay Program. 1991. Report and Recommendation of the Non-Point Source Evaluation Panel, CPB/TRS 56/91. Annapolis, Md.

Chevassus, B. 1979. Hybridization in salmonids: Results and perspectives. Aquaculture 17:113-128.

Chevassus, B. 1983. Hybridization in fish: Aquaculture 33:245-262.

Chew, K.K. 1990. Global bivalve shellfish introductions. Journal of World Aquaculture 21(3):9-22.

Chew, K.K., and D. Toba. 1991. Western region aquaculture industry: Situation and outlook report. Western Regional Aquaculture Consortium, University of Washington, Seattle. 23 pp.

Chieng, C., A. Garcia III, and D. Brune. 1989. Oxidation requirements of a formulated micropulverized feed. Journal of the World Aquaculture Society 20(1):24-29.

Cho, C.Y., H.S. Bayley, and S.J. Slinger. 1974. Partial replacement of herring meal with soybean meal and other changes in a diet for rainbow trout *(Salmo gairdneri)*. J. Fish. Res. Bd. Can. 31:1523-1528.

Chourrout, D. 1984. Pressure-induced retention of second polar body and suppression of first cleavage in rainbow trout: Production of all-triploids, all tetraploids, and heterozygous and homozygous gynogenetics. Aquaculture 35:111-126.

Cicin-Sain, B. 1982. Managing the ocean commons: U.S. marine programs in the seventies and eighties. Marine Technology Society Journal 16(4):16-30.

Cicin-Sain, B., and R.W. Knecht. 1985. The problem of governance of U.S. ocean resources and the new Exclusive Economic Zone. Ocean Development and International Law 15:289.

Cipriano, R.C., J.K. Morrison, and C.E. Starliper. 1983. Immunization of salmonids against *Aaromonas salmonicida*. Journal of World Mariculture Society 14:201-211.

Clardy, G.N., M.J. Fuller, and J.E. Waldrop. 1985. Preliminary Economic Evaluation of Freshwater Shrimp Production in Mississippi. Mississippi Agricultural Economics Research Report 159. 46 pp.

Clarke, R., and M. Beveridge. 1989. Offshore fish farming. Infofish International (3/89):12-15.

Cloern, J.W. 1982. Does the benthos control phytoplankton biomass in South San Francisco Bay? Mar. Ecol. Prog. Ser. 9:191-202.

Coche, A.G. 1984. Aquaculture in marine waters: A list of selected reference books and monographs, 1957-1984. FAO Fisheries Circular No. 723, Revision 2. 29 pp.

Colberg, P. 1978. Effect of ozone on microbial fish pathogens, ammonia, nitrate, nitrite, and BOD in simulated reuse hatchery water. Idaho Agriculture Experiment Station Research Paper No. 7759, Boise.

Colt, J., and R.J. White, eds. 1991. Fisheries Bioengineering Symposium. American Fisheries Society. Bethesda, Md.

Colt, J.E., and D.A. Armstrong. 1981. Nitrogen toxicity to crustaceans, fish and mollusks. In Proceedings of the Bio-Engineering Symposium for Fish Culture, L.J. Allen and E.C. Kinney, eds. Fish Culture Section, American Fisheries Society. Bethesda, Md.

Commission on Marine Science, Engineering and Resources (COMSER). 1969. Our Nation and the Sea: A Plan for National Action. Washington, D.C.: U.S. Government Printing Office.

Conklin, D.E., L.R. D'Abramo, and K. Norman-Boudreau. 1983. Lobster nutrition. Pp. 413-423 in Handbook of Mariculture, Vol. I: Crustacean Aquaculture, J.P. McVey, ed. Boca Raton, Fla.: CRC Press.

Conrad, K.M., M.G. Mast, and J.H. MacNeil. 1988. Performance, yield, and body composition of fingerling channel catfish fed a dried waste egg product. Prog. Fish-Culturist 50:219-224.

Conte, F.S., S.I. Doroshov, P.B. Lutes, and E.M. Strange. 1988. Hatchery Manual for the White Sturgeon *Acipenser transmontanus* Richardson. Cooperative Extension, University of California, Division of Agriculture and National Resources, Pub. 3322. 104 pp.

Cook, M.F., and S.P. Canton. 1988. Calculation of percent gas saturation in water by use of a spreadsheet. Prog. Fish-Culturist 50(4):248-250.

Cook, R.H. 1990. Salmon farming in the Bay of Fundy—The challenge of the future. World Aquaculture 21(2):46.

Council of Economic Advisors. 1989. The Annual Report of the Council of Economic Advisors. Washington, D.C.: U.S. Government Printing Office.

Courtenay, Jr., W.R. Regulation of aquatic invasives in the United States of America with emphasis on fishes. Unpublished manuscript.

Cowery, C.B., J.A. Pope, J.W. Adron, and A. Blair. 1971. Studies on the nutrition of marine flatfish. Growth of the plaice *Plewionectes platessa* on diets containing proteins derived from plants and other sources. Marine Biology 10(2):145-153.

Crook, P.R. 1985. Cage design: Some pointers to cage choice. Fish Farmer 8(6).

Cross, S.F. 1990. Benthic impacts of salmon farming in British Columbia. Report to the British Columbia Ministry of the Environment, Water Management Branch, Victoria, B.C. 78 pp.

Crowder, B., ed. 1982. Sea Grant Aquaculture Plan 1983–1987. Report of the

Aquaculture Committee of the Council of Sea Grant Directors. TAMU-SG-82-114. 37 pp.

Crutchfield, J.A. 1989. Economic aspects of salmon aquaculture. Northwest Environmental Journal 5:37-52.

D'Abramo, L.R., D.E. Conklin, C.E. Bordner, N.A. Baum, and K.A. Norman-Boudreau. 1981. Successful artificial diets for the culture of juvenile lobsters. Journal of the World Mariculture Society 12(1):325-332.

Daniels, H.V., and C.E. Boyd. 1989. Chemical budgets for polyethylene-lined, brackish water ponds. Journal of the World Aquaculture Society 20(2):53-60.

Davies, D. S. 1990. Allocating common property marine resources for coastal aquaculture: A comparative analysis. Ph.D. dissertation, Science Research Center, State University of New York, Stony Brook.

Davis, E.M., G.L. Rumsey, and J.G. Nickum. 1976. Egg-processing wastes as a replacement protein source in salmonid diets. Prog. Fish-Culturist 38:20-22.

DeCrew, M.G. 1972. Antibiotic toxicity, efficacy, and teratogenicity in adult spring chinook salmon (*Oncorhynchus tshawytscha*). J. Fish. Res. Bd. Can. 29(11):1513-1517.

DeVoe, M.R., and A.S. Mount. 1989. An analysis of ten state aquaculture leasing systems: Issues and strategies. Journal of Shellfish Research 8(1):233-239.

Dixon, I. 1986. Fish Farm Surveys in Shetland: Summary and Survey Reports, Vol. 1. A report to NCC, Shetland Islands Council and Shetland Salmon Farmers Assoc. FSC/OPRU/30/80. Orielton Field Center, Pembroke, Dyfed, Scotland.

Dodge, C.H., and W.C. Jolly. 1978. Aquaculture: Status of Technology and Future Prospects. The Library of Congress Congressional Research Service, Issue Brief Number IB77099. 12 pp.

Donaldson, E.M. 1988. Science and the future of aquaculture. Proceedings of the Aquaculture International Congress, Vancouver, B.C. 299-309.

Donaldson, E.M., U.H.M. Fagerlund, D.A. Higgs, and J.R. McBride. 1978. Hormonal enhancement growth. Pp. 456-578 in Fish Physiology, Vol. VIII, W.S. Haar, D.J. Randall, and J.R. Brett, eds. New York: Academic Press.

Doroshov, S.I. 1985. Biology and culture of sturgeon, *Acipenseri forme*. Pp. 251-274 in Recent Advances in Aquaculture, Vol. 2, J.F. Muir and R.J. Roberts, eds. Boulder, Colo.: Westview Press.

Doyle, R.W. 1983. An approach to the quantitative analysis of domestication selection in aquaculture. Aquaculture 33:167-185.

DPA Group. 1988. Cost of production model of pen-rearing of salmon in Alaska and currently producing regions. Report prepared for Alaska Department of Commerce and Economic Development, Juneau.

Dunham, R.A., and R.O. Smitherman. 1983. Crossbreeding channel catfish for improvement of body weight in earthen ponds. Growth 47:97-103.

Dunham, R.A., and R.O. Smitherman. 1983. Response to selection and realized heritability for body weight in three strains of channel catfish, *Ictalurus punctatus*, grown in earthen ponds. Aquaculture 33:89-96.

Dunham, R.A., J. Eash, J. Askins, and T.M. Townes. 1987. Transfer of the metallothionein-human growth hormone fusion gene into channel catfish. Trans. Am. Fish. Soc. 116:87-91.

Duprey, J.L., N.T. Winndsor, and C.E. Sutton. 1977. Manual for Design and Operation of an Oyster Seed Hatchery. Virginia Institute of Marine Science, Gloucester Point, Va.

Dutrieux, E., and O. Guelorget. 1988. Ecological planning: A possible method for the choice of aquacultural sites. Ocean and Shoreline Management 11: 427-447.

Egan, D., and A. Kenney. 1990. Salmon farming in British Columbia. World Aquaculture 21(2):6-11.

Electronic Power Research Institute (EPRI). 1990. A summary description of the second workshop on the role of macroalgal oceanic farming in global change. July 23-24, Newport Beach, California. Electric Power Research Institute, Palo Alto.

Ellis, M. 1990. Decomposition processes on the pond bottom. Presented at Texas Aquaculture Conference, Corpus Christi, February.

Ervik, A., P. Johannessen, and J. Aure. 1985. Environmental Effects of Marine Norwegian Fish Farms. International Council for the Exploration of the Sea, Vol. 37. 13 pp.

Fabi, G., L. Fiorentini, and S. Giannini. 1989. Experimental shellfish culture on an artificial reef in the Adriatic Sea. Bulletin of Marine Science 44(2):923-933.

Fengqi, L. 1990. PRC mariculture update. World Aquaculture 21(2):84-85.

Figueras, A.J. 1989. Mussel aquaculture in Spain and France. World Aquaculture 20(4):8-17.

Fish Farming International. 1990. Alligator earnings boost U.S. farms. 17(5): 21.

Flander-Good Associates. 1989. Economic Assessment of Salmonid Cage Culture Industry in Southwestern New Brunswick. Fredericton, New Brunswick. 105 pp.

Fletcher, G.L., M.A. Shears, M.J. King, P.L. Davies, and C.L. Hew. 1988. Evidence for antifreeze protein gene transfer in Atlantic salmon (*Salmo salar*). Can. J. Fish. Aquat. Sci. 45:352-357.

Folke, C. 1988. Energy economy of salmon aquaculture in the Baltic Sea. Environmental Management 12(4):525-537.

Folsom, W.B., and B.D. McFetters. 1990 World salmon aquaculture. Proceedings of a Marine Technology Society Conference "Science and Technology for a New Oceans Decade," Washington, D.C., 623-628.

Food and Agriculture Organization. 1977. Control of spread of major communicable fish diseases. Report of the FAO/OIE Government Consultation on an International Convention for the Control of the Spread of Major Communicable Fish Diseases. FAO Fisheries Reports No. 192. FID/R192 (EN). FAO/Rome.

Food and Agriculture Organization. 1991. P. 145 in FAO Yearbook, Fishery Statistics, Catches and Landings, Vol. 68. Rome.

Ford, S.E., A.J. Figueras, and H.H. Haskin. 1990. Influence of selective breeding, geographic origin, and disease on gametogenesis and sex ratios of oysters, *Crassostrea virginica*, exposed to the parasite *Haplasporidium nelsoni* (MSX). Aquaculture 88(3/4):285-301.

Friars, G.W., J.K. Bailey, and K.A. Coombs. 1990. Correlated responses to selection for grilse length in Atlantic salmon. Aquaculture 85:171-176.

Fridley, R.B., R.H. Piedrahita, and T.M. Losordo. 1988. Challenges in aquacultural engineering. Agricultural Engineering (May/June):12-15.

Gall, G.A.E. 1990. Basis for evaluating breeding plans. Aquaculture 85:125-142.

Getchell, R. 1988. Environmental effects of salmon farming. Aquatic Magazine (November/December):44-47.

Gillespie, D. 1986. An inquiry into finfish aquaculture in British Columbia: Report and recommendations. Prepared for Government of British Columbia, December. 50 pp.

Gjerdem, T. 1983. Genetic variation in quantitative traits and selection breeding in fish and shellfish. Aquaculture 33:51-72.

Goswami, V., S.G. Dalal, and S.C. Goswami. 1986. Preliminary studies on prawn, *Penaeus merguiensis*, for selection of broodstock in genetic improvement programs. Aquaculture 53(1):41-48.

Gousset, G. 1990. European eel (*Anguilla* L.) farming technologies in Europe and in Japan: Application of a comparative analysis. Aquaculture 87:209-235.

Gowen, R.J. 1988. Release strategies for coho and chinook salmon released into Coos Bay, Oregon. In Salmon Production, Management, and Allocation—Biological, Economic and Policy Issues, William J. McNeil, ed. Oregon State University Press.

Gowen, R.J., and N.B. Bradbury. 1987. The ecological impact of salmon farming in coastal waters: A review. Oceanography and Mar. Biol. Annual Rev. 25:562-575.

Gowen, R.J., and D.S. McLusky. 1990. Investigation into Benthic Enrichment, Hypernutrification and Eutrophication Associated with Mariculture in Scottish Coastal Waters. Summary of main report to Highlands and Islands Development Board, Crown Estate Comm., Countryside Comm. for Scotland. National CONS. Council and Scottish Salmon Growers Association 13 pp.

Gowen, R.J., N.B. Bradbury, and J.R. Brown. 1985. The Ecological Impact of Salmon Farming in Scottish Coastal Waters: A Preliminary Appraisal. International Council for the Exploration of the Sea, Vol. 35, 13 pp.

Grados, O.G. 1991. Diet and seasonal nutrient variation in earthen ponds stocked with *Penaeus vannamei* at three population densities. M.S. thesis, College of Charleston, Charleston, S.C., 95 pp.

Grant, B.F., P.A. Seib, M. Liao, and K.E. Corpron. 1989. Polyphosphorylated L-ascorbic acid: A stable form of vitamin C for aquaculture feeds. Journal of the World Aquaculture Society 20(3):143-157.

Griffiths et al. 1991. Distribution and dispersal of the zebra mussel *(Dreissena polymorpha)* in the Great Lakes Region. Canadian J. of Fisheries and Aquatic Sciences 48:1381-1388.

Grove, R.S., C.J. Sonu, and M. Nakumura. 1989. Recent Japanese trends in fishing reef design and planning. Bulletin of Marine Science 44(2):984-996.

Guiry, M.D. 1984. Structure, life history, and hybridization of Atlantic *Gigartina teedii* (Rhodophyta) in culture. Br. Phycol. J. 19(1):37-55.

Gulf States Marine Fisheries Commission. 1990. Summary of aquaculture programs by state. A report to the Technical Coordinating Committee, Orange Beach, Ala. March 14.

Gulland, J.A. 1971. The Fish Resources of the Ocean. Surrey: Fishing News Books Ltd., 255 pp.

Haby, M. 1990. Marketing opportunities for red drum. Pp. 209-213 in Red Drum Aquaculture, Chamberlain, G., R. Miget, and M. Haby, eds. Texas A&M University. College Station.

Hagood, R.W., G.N. Rothwell, M. Swafford, and M. Tosaki. 1981. Preliminary report on the aquaculture development of the dolphin fish, *Coryphaena hippurus* (Linnaeus). Journal of the World Mariculture Society 12(1):135-139.

Hallerman, E.M., and A.R. Kapucinski. 1990. Transgenic fish and public policy: Regulatory concerns. Fisheries (Bethesda) 15 (1):12-20.

Hanfman, D.T. 1987. Aquaculture: Development Plans and Marketing. Quick Bibliography Series. U.S. Department of Agriculture, Washington, D.C.

Hanfman, D.T. 1987. Shellfish Culture, 1979-1986. Quick Bibliography Series. U.S. Department of Agriculture, Washington, D.C. 15 pp.

Hanfman, D.T. 1988. Shrimp Mariculture, 1979-1987. Quick Bibliography Series. U.S. Department of Agriculture, Washington, D.C. 7 pp.

Hanfman, D.T. 1988. Water Quality Management. Quick Bibliography Series. U.S. Department of Agriculture, Washington, D.C. 7 pp.

Hanfman, D.T. 1989. Salmon Culture, January 1979-September 1988. Quick Bibliography Series. U.S. Department of Agriculture, Washington, D.C. 14 pp.

Hanfman, D.T. 1990. Aquaculture Development and Economics, January-December 1990. Quick Bibliography Series 90-41. U.S. Department of Agriculture, Washington, D.C.

Hanfman, D.T., S. Tibbitt, C. Watts, and D. Alston. 1988. Aquaculture in the Caribbean Basin: A Bibliography (1970-88). Prepared by the U.S. Aquatic Sciences and Fisheries Abstracts (ASFA), 71 pp.

Hanfman, D.T., S. Tibbitt, and C. Watts. 1989. The Potentials of Aquaculture: An Overview and Bibliography. Bibliographies and Literature of Agriculture No. 90. U.S. Department of Agriculture. Washington, D.C.

Hansen, J.E., J.E. Packard, and W.T. Doyle. Mariculture of Red Seaweeds. California Sea Grant College Program Publication, Report No. T-CSGCP-002.

Harkness, W.J.K., and J.R. Dymond. 1961. The Lake Sturgeon, the History of Its Fishery and Problems of Conservation. Ontario Department of Lands Forests, Toronto, 121 pp.

Harvey, B.J. 1987. Gamete banking and applied genetics in aquaculture. In Selection, Hybridization and Genetic Engineering in Aquaculture, Vol. 1, K. Tiews, ed. Proceedings of the World Symposium on Selection, Hybridization and Genetic Engineering. Bordeaux, France. June 27-30, 1986.

Heap, S.P., and J.P. Thorpe. 1987. A preliminary study of comparative growth rates in O-group malpigmented and normally pigmented turbot, *Scophthalmus maximus* (L.) and turbot-brill hybrids, *S. maximus* X *S. rhombus* (L.) at two temperatures. Aquaculture 60:251-264.

Hedgecock, D., and S.R. Malecha. 1990. Prospects for the application of biotechnology to the development of shrimp and prawns. In Shrimp Culture in North America and the Caribbean, P.A. Sandifer, ed. Advances in World Aquaculture, World Aquaculture Society, Baton Rouge, La.

Helm, L. 1989. Trouble down on the fish farm. *Seattle Post-Intelligencer*. (Dec. 18), pp. B1, B4.

Herman, R.L., D. Collis, and G.L. Bullock. 1969. Oxytetracycline residues in different tissues of trout. Bureau of Sport Fishing and Wildlife. Technical Paper No. 37. U.S. Department of the Interior, Washington, D.C.

Hershberger, W.K., J.M. Myers, R.N. Iwamoto, W.C. McAuley, and A.M. Saxton.

1990. Genetic changes in the growth of coho salmon *(Oncorhynchus kisutch)* in marine net-pens produced by ten years of selection. Aquaculture 85:187-198.

Hetrick, J. 1990. Salmon ranching—Alaskan gold. Canadian Aquaculture 6(2):33-35.

Hetrick, J. 1991. Alaskan aquaculture. *Water Farming Journal* 66(4):10-13 (April).

Hew, C.L. 1989. Transgenic fish: Present status and future directions. Fish Physiology and Biochemistry 7(104):409-413.

Hew, C.L., P.L. Davies, M.A. Shears, M.J. King, and G.L. Fletcher. 1987. Antifreeze protein gene transfer to Atlantic salmon by micro-injection. Fed. Proc. 46:2039.

Hicks, B. 1989. Fish health regulations restrict industry not disease. Canadian Aquaculture 6(1):27-28.

Hindar, K., N. Ryman, and F. Utter. 1991. Genetic effects of aquaculture on natural fish populations. Can. J. Fish. Aquat. Sci. 48:945-957.

Hinshaw, R.N. 1973. Pollution as a Result of Fish Cultural Activities. EPA Report EPA-R3-73-009, National Technical Information Service PB-221-376. 209 pp.

Hjul, Peter. 1973. FAO conference on fishery management and development. Fishing News International (May):20-35.

Hoffman, G.L. 1970. Intercontinental and transcontinental dissemination and transfaunation of fish parasites with emphasis on whirling disease (*Myxosoma cerebralis)*. Pp. 69-81 in A symposium on Diseases of Fish and Shellfishes, S.F. Sniezko, ed. American Fisheries Society Special Publication 5.

Hollerman, W.D., and C.E. Boyd. 1985. Effects of annual draining on water quality and production of channel catfish in ponds. Aquaculture, 46:45-54.

Holt, G.J. 1990. Growth and development of red drum eggs and larvae. Pp. 46-50 in Red Drum Aquaculture, Texas A&M Sea Grant College Program, College Station. TAMU-SG-90-603.

Holt, G.J. 1991. Intensive culture of larval red drum: experimental studies. Journal of the World Aquaculture Society.

Holt, G.J. 1992. Experimental studies of feeding of larval red drum. Journal of the World Aquaculture Society. (In press.)

Homziak, J., and J.D. Lunz, eds. 1983. Aquaculture in Dredged Material Containment Areas; Proceedings. Environmental Laboratory, U.S. Army Engineer Waterways Experiment Station, Miscellaneous Paper D-83-2. 216 pp.

Hopkins, J.S. 1991. Status and history of marine and freshwater shrimp farming in South Carolina and Florida. Pp. 17-35 in Shrimp Culture in North America and the Caribbean, P.A. Sandifer, ed. The World Aquaculture Society, Baton Rouge, La.

Hopkins, J.S., M.L. Baird, O.G. Grados, P.P. Maier, P.A. Sandifer, and A.D. Stokes. 1988. Impact of intensive shrimp production on the culture pond ecosystem. Journal of the World Aquaculture Society 19 (1):37A(abstract).

Hopkins, J.S., M.L. Baird, O.G. Grados, P.P. Maier, P.A. Sandifer, and A.D. Stokes. 1988. Impacts of Intensive Shrimp Culture Practices on the Culture Pond Ecology. Report from the Waddell Mariculture Center of the South Carolina Marine Resources Division.

Hopkins, J.S., A.D. Stokes, C.L. Browdy, and P.A. Sandifer. Submitted. The relationship between feeding rate, paddlewheel aeration rate and expected dawn dissolved oxygen in intensive shrimp ponds. Aquacultural Engineering.

Hornstein, D. 1980. Salmon Ranching in Oregon: State and Federal Regulations. Oregon State University Extension Marine Advisory Program, Special Report 573. 8 pp.

Huang, H., W.L. Griffin, and D. Aldrich. 1984. A preliminary economic feasibility analysis of a proposed commercial penaeid shrimp culture operation. Journal of the World Mariculture Society 15:95-105.

Hughes, J.T., J.J. Sullivan, and R. Shleser. 1972. Enhancement of lobster growth. Science 177:1110-1111.

Huguenin, J.E., and J. Colt. 1989. Design and Operating Guide for Aquaculture Seawater Systems. Developments in Aquaculture and Fisheries Science, 20. New York: Elsevier. 264 pp.

Hunter, G.A., E.M. Donaldson, J. Stoss, and I. Baker. 1983. Production of monosex female groups of chinook salmon *(Oncorhynchus tshawytscha)* by the fertilization of normal ova with sperm from sex-reversed females. Aquaculture 33:355-364.

Hymel, T.M. 1985. Water quality dynamics in commercial crawfish ponds and toxicity of selected water quality variables to *Procambarus clarkii*. M.S. thesis for the Louisiana State University School of Forestry, Wildlife and Fisheries, Baton Rouge. 119 pp.

Ihssen, P.E., L.R. McKay, I. McMillan, and R.B. Phillips. 1990. Ploidy manipulation and gynogenesis. Fishes: Cytogenetic and Fisheries Applications, Transactions of the American Fisheries Society 119(4):698-717.

Imai, T., ed. 1977. Aquaculture in Shallow Seas: Progress in Shallow Sea Culture (translated from Japanese). National Oceanic and Atmospheric Administration, Washington, D.C. 615 pp.

Industry Task Force on Aquaculture. 1984. Aquaculture: A Development Plan for Canada. Final Report of the Industry Task Force on Aquaculture sponsored by the Science Council of Canada, Ottowa, Ontario. 22 pp.

Institute of Medicine. 1991. Seafood Safety. Washington, D.C.: National Academy Press.

Institution of Civil Engineers. 1990. Proceedings of the Conference on Engineering for Offshore Fish Farming. London: Thomas Telford.

International Council for Exploration of the Sea (ICES). 1984. Guidelines for implementing the ICES code of practice concerning introductions and transfers of marine species. Cooperative Research Report 130, 1-20.

International Council for the Exploration of the Sea (ICES). 1987. Report of the ad hoc study group on environmental impacts of mariculture. Copenhagen: ICES Cooperative Research Report 164. 83 pp.

International Council for the Exploration of the Sea (ICES). 1988. Report of the ad hoc study group on environmental impacts of mariculture. Copenhagen: ICES Cooperative Research Report 164. 83 pp.

International Council for the Exploration of the Sea (ICES). 1988. Report of the Working Group on Environmental Impacts of Mariculture. April. Hamburg, Germany. 1989/F:11, 70 pp.

Isaacs, F. 1990. Irish moss aquaculture moves from lab to marketplace. World Aquaculture 21(2):95-97.

Israel, D., F. Apud, and N. Franco. 1984. The economics of different prawn and shrimp pond culture systems: A comparative analysis. In Proceedings of the First International Conference on the Culture of Penaeid Prawns/Shrimps, Y. Taki, J.H. Primavera, and J.A. Llobrera, eds. Iloilo City, Philippines, December 4-7.

Itami, T., Y. Takahashi, and Y. Nakamura. 1989. Efficacy of vaccination against vibriosis in cultured Kuruma prawns *Penaeus japonicus*. Journal of Aquatic Animal Health 1:238-242.

Iwamoto, R.N., J.M. Myers, and W.K. Hershberger. 1986. Genotype-environment interactions for growth of rainbow trout, *Salmo giardneri*. Aquaculture 57(104): 153-162.

Jacobsen, M.D. 1989. Withdrawal times of freshwater rainbow trout, *Salmo giardneri*, after treatment with oxolinic acid, oxytetracycline and trimetoprim. J. Fish Diseases 12:29-36.

Jahncke, M.L., M.B. Hale, J.A. Gooch, and J.S. Hopkins. 1988. Comparison of pond-raised wild red drum *(Sciaenops ocellatus)* with respect to proximate composition, fatty acid profiles, and sensory evaluations. J. Food Sci. 53(1):286-287.

Jahncke, M.L., T.I.J. Smith, and G.T. Seaborn. 1988. Use of fatty acid profiles to distinguish cultured from wild fish: A possible law enforcement tool. Annual Conference of the Southeastern Association of Fisheries and Wildlife Agency.

Jahncke, M.L., T.I.J. Smith, and G.T. Seaborn. 1989. Fatty acid profiles: A potential method to differentiate wild from cultured fish. Northwest Association of Forensic Scientists (abstract). Ashland, Ore.: Spring 1989.

Japanese Ministry of Agriculture, Forestry and Fishery. 1990. 1989 Update on Domestic Production. Statistics and Information Bureau (Gyogyo Yoshokugyo Seisan Tokei Nempo), Tokyo, Japan.

Jobling, M. 1988. A review of the physiological and nutritional energetics of cod, *Gadus morhua* L., with particular reference to growth under farmed conditions. Aquaculture 70:1-19.

Johnstone, R. 1985. Induction of triploidy in Atlantic salmon by heat shock. Aquaculture 49:133-139.

Joint Subcommittee on Aquaculture. 1990. National Aquaculture Forum Output, November 1987, draft of September 7, 1990.

Joint Subcommittee on Aquaculture. 1991. Meeting minutes for September 12, 1990, December 18, 1991, and April 12, 1991, provided by R.O. Smitherman.

Jonas, M., J.B. Comer, and B.A. Cunha. 1984. Tetracyclines. In Antimicrobial Therapy, A.M. Ristuccia and B.A. Cunha, eds. New York: Raven Press.

Juan, Y., W.L. Griffin, and A.L. Lawrence. 1988. Production costs of juvenile penaeid shrimp in an intensive greenhouse raceway nursery system. Journal of the World Aquaculture Society 19(3):149-160.

Juario, J.V., and L.V. Benitez, eds. 1988. Perspectives in Aquaculture Development in Southeast Asia and Japan. Aquaculture Department SEAFDEC, Tigbauan, Iloilo, Philippines.

Kaiser, G.E., and F.W. Wheaton. 1991. Engineering aspects of water quality monitoring and control. Pp. 210-232 in Engineering Aspects of Intensive Aquaculture. Proceedings from the Aquaculture Symposium. Northeast Regional Agricultural Engineering Service, Ithaca, N.Y.

Kajikawa, T., H. Takazawa, M. Amano, A. Murata, K. Kitani, and H. Kitano. 1988. An ocean-based mariculture-OTEC system. Paper presented at Pacific Congress on Marine Science and Technology (PACON 88). 7 pp.

Kalagayan, G., D. Godin, R. Kanna, G. Hagino, J. Sweeney, and J. Wyban. 1990. IHHN virus as an etiological factor in runt-deformity syndrome of juvenile *Penaeus vannamei* cultured in Hawaii. World Aquaculture Society, World Aquaculture 90 (abstract T17.2).

Kanazawa, A., S. Koshio, and S. Teshina. 1989. Growth and survival of larval red sea bream, *Pagrus major* and Japanese flounder, *Paralichthys olivaceus* fed microbound diets. Journal of the World Aquaculture Society 20(2):31-37.

Kantrowitz, B.M. 1984. Launching a seaweed farm, a future source of energy. Aquaculture 11(2):29-31.

Kapuscinski, A.R. and E.M. Hallerman. 1990. Transgenic fish and public policy: Anticipating environmental impacts of transgenic fish. Fisheries (Bethesda) 15(1):2-11.

Kapuscinski, A.R., and E.M. Hallerman. 1990. Transgenic fishes: AFS position statement. Fisheries (Bethesda) 15(4):2.5.

Kearney/Centaur. 1988. Development of value added, margin and expenditures for marine fishery products. Prepared for U.S. Department of Commerce, National Marine Fisheries Service, Washington, DC.

Keen, E. 1988. Ownership and Productivity of Marine Fishery Resources. Blacksburg, Va.: The McDonald and Woodward Publishing Company.

Keller, S. 1987. Proceedings of the Fourth Alaska Aquaculture Conference. Alaska Sea Grant Report No. 88-4. 18 pp.

Kerr, N.M., M.J. Gillespie, S.T. Hull, and S.J. Kingwell. 1980. The design, construction, and location of marine floating cages. Pp. 70-83 in Proceedings of the Institute of Fisheries Management Cage Fish Rearing Symposium, University of Reading, London. Janssen Services.

Ketola, H.G. 1975. Requirement of Atlantic salmon for dietary phosphorus. Trans. Amer. Fish. Soc. 104(3):543-551.

Ketola, H.G. 1982. Effect of phosphorus in trout diets on water pollution. Salmonid 6(2):12-15.

Ketola, H.G. 1985. Mineral nutrition: Effects of phosphorus in trout and salmon feeds on water pollution. Pp. 465-473 In Nutrition and Feeding of Fish, C.B. Cowery, A.M. Mackie, and J.G. Bell, eds. London: Academic Press.

Ketola, H.G. 1988. Salmon fed low-pollution diet thrive in Lake Michigan. U.S. Fish and Wildlife Service, Research Information Bulletin No. 62-25, April.

Ketola, H.G. 1990. Studies on diet and phosphorus discharges in hatchery effluents (abstract). International NSMAW Symposium. Guelph, Canada, June 5-9, 1990.

Ketola, H.G., H. Westers, C. Pacor, W. Houghton, and L. Wubbels. 1985. Pollution: Lowering levels of phosphorus, experimenting with feed. Salmonid 9(2):11.

Ketola, H.G., M. Westers, W. Houghton, and C. Pecor. 1990. Effects of diet on growth and survival of coho salmon and on phosphorus discharges from a fish hatchery. American Fisheries Society Symposium No. 11.

Kincard, A.L., W.R. Bridges, and B. Von Limbach. 1977. Three generations of selection for growth rate in fall-spawning rainbow trout. Trans. Am. Fish. Soc. 106:621-628.

King, S.T., and J.R. Schrock. 1985. Controlled Wildlife, A Three Volume Guide to U.S. Wildlife Laws and Permit Procedures. III. State Wildlife Regulations. Association of Systematics Collections, Lawrence, Kan.

Kinghorn, B.P. 1983. A review of quantitative genetics in fish breeding. Aquaculture 31(2,3,4):283-304.

Klontz, G.W., I.R. Brock, and J.A. McNair. 1978. Aquaculture Techniques: Water Use and Discharge Quality. Idaho Water Resource Research Institute, National Technical Information Service report PB-285-956, 114 pp.

Kneale, D.C., S.E. Sulman, D.K. Roberts, and H.J. Khalafalla. 1982. Studies on greenhouse temperature control and water flow for intensive shrimp culture. Annual Res. Rep. Kuwait Inst. Sci. Res. 1981:96-99.

Knecht, R.W. 1986. In Ocean Resources and U.S. Intergovernmental Relations, M. Silva, ed. Boulder, Colo.:Westview Press.

Knecht, R.W., B. Cicin-Sain, and J.H. Archer. 1988. National ocean policy: A window of opportunity. Ocean Development and International Law 19:113-142.

Korringa, P. 1976. Farming Cupped Oysters of the Genus *Crassostrea*. A Multidisciplinary Treatise. Amsterdam: Elsevier. 224 pp.

Korringa, P. 1976. Farming Flat Oysters of the Genus *Ostrea*. A Multidisciplinary Treatise. Amsterdam: Elsevier. 238 pp.

Krom, M.D., C. Porter, and H. Gordin. 1985. Nutrient budget of a marine fish pond in Eilat, Israel. Aquaculture 51: 65-80.

Kruner, G., and H. Rosenthal. 1983. Efficiency of nitrification in trickling filters using different substrates. Aquacultural Engineering 2:49-67.

Kvalheim, H. 1988. Profitability of tomorrow's salmon farming. Proceedings of the Aquaculture International Congress, Vancouver, B.C. 233-239.

Landless, P. 1985. Aeration in floating cages. Fish Farmer 8(3).

Lannan, J.E., and A.R. Kapuscinski. 1986. Application of a genetic fitness model to extensive aquaculture. Aquaculture 57:81-87.

Lannan, J.E., R.O. Smitherman, and G. Tchobanoglas, eds. 1985. Principles and Practices of Pond Aquaculture. Oregon State University Press, Corvallis.

Larsson, A.M. 1985 Blue mussel sea farming—Effects on water quality. Vatten 41: 218-224.

Lawrence, A.L., G.W. Chamberlain, and D.L. Hutchins. 1981. Shrimp Mariculture. Shrimp Mariculture Project, The Texas A&M University Sea Grant College Program. 9 pp.

Leary, R.F., F.W. Allendorf, K.L. Knudsen, and G.H. Thorgaard. 1985. Heterozygosity and developmental stability in gynogenetic diploid and triploid rainbow trout. Heredity 54:219-225.

Leighton, D.L. 1985. Rock scallop growout. Aquaculture 11(4):6-8.

Leonard, D.L., and E.A. Slaughter. 1990. The quality of shellfish growing waters on the West Coast of the United States. U.S. Department of Commerce, National Oceanic and Atmospheric Administration, National Estuarine Inventory. 51 pp.

Lester, L.J. 1983. Developing a selective breeding program for penaeid shrimp mariculture. Aquaculture 33:41-50.

Liao, P.B., and R.D. Mayo. 1972. Salmonid hatchery water reuse systems. Aquaculture 1: 317-335.

Liao, P.B., and R.D. Mayo. 1974. Intensified fish culture combining water reconditioning with pollution abatement. Aquaculture 3:61-85.

Lim, C., and W. Domininy. 1990. Evaluation of soybean meal as a replacement for marine animal protein in diets for shrimp *(Penaeus vannamei)*. Aquaculture 87: 53-63.

Linfoot, B.T., and M.S. Hall. 1987. Analysis of the motions of scale-model, seacage systems. In Automation and Data Processing in Aquaculture, J.G. Balchen and A. Tysso, eds. IFAC Proceedings 1987, No. 9. Pergamon Press.

Losordo, T.M. 1991. Engineering consideration in closed recirculating systems. Pp. 58-69 in Aquaculture Systems Engineering. Proceedings of the World Aquaculture Society and the American Society of Agricultural Engineers Jointly Sponsored Session at the World Aquaculture Society Meeting, San Juan, Puerto Rico.

Losordo, T.M., R.H. Piedrahita, and J.M. Ebeling. 1988. An automated data acquisition system for use in aquaculture ponds. Aquacultural Engineering 7:265-278.

Love, R.M. 1987. Stress and behavior in the culture environment. Pp. 449-472 in Realism in Aquaculture: Achievements, Constraints, Perspectives, M. Bilio, H. Rosenthal, and C.J. Sindermann, eds. European Aquaculture Society, Bredene, Belgium.

Lovell, T. 1989. Nutrition and Feeding of Fish. New York: Van Nostrand Reinhold. 260 pp.

Lutz, R.A. 1985. Mussel aquaculture in the United States. Pp. 311-363 in Crustacean and Mollusk Aquaculture in the United States, J.V. Huner and E.E. Brown, eds. Westport, Conn.: AVI Publishing Co.

MacCrimmon, H.R. 1971. World distribution of rainbow trout (*Salmo giardneri*). J. Fish. Res. Bd. Can. 28(5):663-704.

Maclean, J.T. 1988. Nonpoint Source Pollution January 1984-August 1988. Quick Bibliography Series U.S. Department of Agriculture, Washington D.C. 18 pp.

Maclean, N., and D. Penman. 1990. The application of gene manipulation to aquaculture. Aquaculture 85:1-20.

Main, K.L., and E. Antill. 1988. Warmwater aquaculture industry target of development. Aquatic Magazine (July/August):20-25.

Maine, State Planning Office. 1987. Establishing the Maine Advantage: An Economic Development Strategy for the State of Maine.

Malone, R.F., and D.G. Burden. 1988. Design of recirculating blue crab shedding systems. Louisiana Sea Grant College Program, Center for Wetland Research, Louisiana State University, Baton Rouge.

Malouf, R.E. 1989. Clam culture as a resource management tool. Pp. 427-447 in Clam Mariculture in North America, J.J. Manzi and M. Castagna, eds. Amsterdam: Elsevier.

Manci, W.E. 1990. Private consultants in aquaculture and information transfer. Fisheries (Bethesda) 15(4):6-10.

Manzi, J.J. 1985. Clam aquaculture. Pp. 275-310 in Crustacean and Mollusk Aquaculture in the United States, J.V. Huner and E.E. Brown, eds. Westport, Conn.: AVI Publishing Co.

Manzi, J.J. 1990. The role of aquaculture in the restoration and enhancement of molluscan fisheries in North America. Pp. 53-56 in Marine Farming and

Enhancement, A.K. Sparks, ed. Proceedings of the 15th U.S.-Japan Meeting on Aquaculture, Kyoto, Japan, October 22-23. NOAA Technical Report NMFS 85.

Marcus, J.M. 1985. A Special Water Quality Sampling of Three Crawfish Aquaculture Systems Sumter and Georgetown Counties, South Carolina. South Carolina Department of Health and Environmental Control, Office of Environmental Control, Technical Report 032-84. 11 pp.

Marking, L.L., G.E. Howe, and J.R. Crowther. 1988. Toxicity of erythromycin, oxytetracycline, and tetracycline administered to lake trout in water baths by injection or by feeding. Prog. Fish-Culturist 50:197-201.

Martin, J.L., and A. White. 1988. Distribution and abundance of the toxic inflagellate *Gonyaulax excavata* in the Bay of Fundy. Can. J. Fisheries Aquat. Sci. 45:1968-1975.

Martin, M. 1983. Goldfish farming. Aquatic Magazine 9(3): 38-40; 9(4): 38-40; 9(5): 30-34.

Mathisen, O.A., and T. Gudjonsson. 1978. Salmon management and ocean ranching in Iceland. J. Agr. Res. Iceland 10(2):156-174.

Mayo Associates. 1986. Planning report amended for the Parkview Hatchery Reconstruction Project. Prepared for the State of New Mexico Department of Game and Fish (with Leedshill-Herkenhoff, Inc.).

Mayo Associates. 1987. A facility development program for SilverKing Oceanic Farms of Santa Cruz, California.

Mayo Associates. 1988a. An Assessment of Private Salmon Ranching in Oregon. Prepared for the Oregon Coastal Zone Management Association, Seattle, Wash. 85+ pp.

Mayo, R.D. 1976. A technical and economic review of the use of reconditioned water in aquaculture. Presented at FAO Conference on Aquaculture, Kyoto, Japan, May 26–June 2.

Mayo, R.D. 1981. Recirculation Systems in Northern America. Prepared for the World Symposium on Aquaculture in Heated Effluents and Recirculation Systems. Stavanger, May 28-30, 1980, Vol. II. Berlin.

Mayo, R.D. 1988. The Bio-Engineer's Waterproof, Expandable Handbook of Factors, Prices and Concepts Needed While Trying to do a Feasibility Study for a Shrimp Farm (or Whatever) in an Unheated Room, Late at Night, in Bird Song Junction, Anywhere. Prepared for the World Aquaculture Society, to be presented in Hawaii, January 5-9.

Mayo, R.D. 1988. The Bird Song Junction Handbook. Prepared for the 1988 Annual Meeting of the World Aquaculture Society. The Mayo Associates, Seattle, Wash. 100 pp.

Mayo, R.D. 1989. A review of water reuse. World Aquaculture Society Annual Meeting, Los Angeles.

Mayo, R.D., C.M. Brown, J. Colt, and J. Glude. 1989. The California hatchery evaluation study. World Aquaculture Society Annual Meeting, Los Angeles, February 12-16. Washington, D.C.: U.S. Government Printing Office,

Mazzaccaro, T. 1988. Aquaculture Production Systems Other Than Ponds 1979-1987. Quick Bibliography Series, U.S. Department of Agriculture, Washington, D.C. 12 pp.

McCarty, C.E., J.G. Geiger, L.N. Sturmer, B.A. Gregg, and W.P. Rutledge. 1986.

Marine finfish culture in Texas: A model for the future. Pp. 249-262 in Fish Culture in Fish Management, R.H. Stroud, ed. American Fisheries Society, Washington, D.C.

McDonald, C.D., and H.E. Deese. 1988. Hawaii's ocean industries: Relative economic status. Proc. PACON 88, Pacific Cong. Mar. Sci. Technol. 16-20.

McEvoy, T., M. Stack, B. Beane, T. Barn, J. Srednan, and F. Gannon. 1988. The expression of a foreign gene in salmon embryos. Aquaculture 68:27-37.

McGee, M.V., and C.E. Boyd. 1983. Evaluation of the influence of water exchange in channel catfish ponds. Trans. Amer. Fish. Soc. 112:557-560.

McLaughlin, T.W. 1981. Hatchery effluent treatment—U.S. Fish and Wildlife Service. In Proceedings of the Bio-Engineering Symposium for Fish Culture, L.J. Allen and E.C. Kinney, eds. Fish Culture Section, American Fisheries Society, Bethesda, Md.

McLean, E., and R. Ash. 1990. Modified uptake of the protein antigen horseradish peroxidase (HRP), following oral delivery of rainbow trout, *Oncorhynchus mykiss*. Aquaculture 87(3/4):373-380.

McNeil, W.J. 1980. Salmon ranching in Alaska. Pp. 13-27 in Salmon Ranching, J. E. Thorpe, ed. New York: Academic Press.

McNeil, W.J. 1988. Salmon Production, Management, and Allocation—Biological, Economic, and Policy Issues, W.J. McNeil, ed. Oregon State University Press.

McVey, E.M. 1989. Aquaculture for Youth and Youth Educators. Aqua-Topics, U.S. Department of Agriculture, Washington, D.C.

McVey, E.M. 1990. Aquaculture in Recirculating Systems, January 1979–December 1989. Quick Bibliography Series 90-26, U.S. Department of Agriculture, Washington, D.C.

McVey, E.M. 1990. Shrimp Mariculture, January 1979-January 1990. Quick Bibliography Series 90-38, U.S. Department of Agriculture, Washington, D.C.

Meltzoff, S., and E. LiPuma. 1985/6. The Social Economy of Coastal Resources: Shrimp Mariculture in Ecuador. Culture and Agriculture (Winter 85/86).

Meriwether II, F.H., E.D. Scura, and W.Y. Okamura. 1983. Culture of red tilapia in freshwater prawn and brackish water ponds. Pp. 260-267 in Proceedings 1st International Conference on Warm Water Crustacea, Brigham Young University, Laie, Hawaii.

Meriwether II, F.H., E.D. Scura, and W.Y. Okamura. 1984. Cage culture of red tilapia in prawn and shrimp ponds. Journal of the World Aquaculture Society 15:254-265.

Miller, G.E., and G.S. Libey. 1985. Evaluation of three biological filters suitable for aquacultural applications. Journal of the World Mariculture Society 16:158-168.

Mirza, J., and W.L. Shelton. 1988. Induction of gynogenesis and sex reversal in silver carp. Aquaculture 68:1-14.

Moss, R.R., and M.S. Doty. 1987. Establishing a Seaweed Industry in Hawaii: An Initial Assessment. A study produced for the Aquaculture Development Program of the Hawaii State Department of Land and Natural Resources. 73 pp.

Mudrak, V.A. 1981. Guidelines for economical commercial fish hatchery wastewater treatment systems. In Proceedings of the Bio-Engineering Symposium for Fish Culture, L.J. Allen and E.C. Kinney, eds. Fish Culture Section, American Fisheries Society, Bethesda, Md.

Muir, J.F. 1985. Aquaculture—Towards the future. Endeavour, New Series 9(1): 52-55.

Muise, B. 1990. Mussel culture in Eastern Canada. World Aquatics 21(2):12-23.

Munro, A.S.S., and I.F. Wadell. 1984. Furunculosis: Experience of Its Control in the Sea Water Cage Culture of Atlantic Salmon in Scotland. International Council for the Exploration of the Sea CM/F:32. 9 pp.

Murawski, S.A., and F.M. Serchuk. 1989. Environmental Effects of Offshore Dredge Fisheries for Bivalves. Prepared for the ICES 1989 Statutory Meeting. Shellfish Committee 1989/K:27. 19 pp.

Naef, F.E. 1971. Pan-size salmon from ocean systems. Sea Grant 70's 2(4):1-2.

Nash, C.E. 1987. Future Economic Outlook for Aquaculture and Related Assistance Needs. Aquaculture Development and Coordination Programme, Food and Agriculture Organization, ADCP/REP/87/25.

Nash, C.E. 1988. Aquaculture communiques: A global overview of aquaculture production. Journal of the World Aquaculture Society 19(2):51-58.

Nash, C.E., and C.B. Kensler. 1990. A global overview of aquaculture production in 1987. World Aquaculture 21(2):104-112.

National Academy of Sciences. 1978. Aquaculture in the United States: Constraints and Opportunities. Washington, D.C.: National Academy Press.

National Academy of Sciences. 1980. The Effects on Human Health of Subtherapeutic Use of Antimicrobials in Animal Feeds. Committee to Study the Subtherapeutic Antibiotic Use in Animal Feeds. Washington, D.C.: National Research Council.

National Aquaculture Development Plan, Vol. I. 1981. Prepared by the Joint Subcommittee of the Federal Coordinating Council on Science, Engineering, and Technology, Washington, D.C.: U.S. Government Printing Office. 67 pp.

National Aquaculture Development Plan, Vol. II. 1983. Prepared by the Joint Subcommittee on Aquaculture of the Federal Coordinating Council on Science, Engineering, and Technology, Washington, D.C.: U.S. Government Printing Office.

National Fisherman. 1991. Suit over offshore salmon farm plan. 72:2(June)8.

National Oceanic and Atmospheric Administration. 1977. NOAA Aquaculture Plan. Washington, D.C.: U.S. Government Printing Office.

National Research Council. 1974. Nutrient Requirements of Trout, Salmon, and Catfish. Board on Agriculture and Renewable Resources, Washington, D.C.: National Academy Press.

National Research Council. 1974. Research Needs in Animal Nutrition. Board on Agriculture and Renewable Resources. Washington, D.C.: National Academy Press.

National Research Council. 1977. World Food and Nutrition Study: Panel on Aquatic Food Resources. Commission on International Relations. Washington, D.C.: National Academy Press.

National Science Foundation. 1991. Workshop on Engineering Research Needs for Off-Shore Mariculture Systems. East-West Center, University of Hawaii. September 26-28.

Needham, T. 1990. Canadian aquaculture—Let's farm the oceans. World Aquaculture 21(2):76-80.

Nehlsen, W., E. Williams, and J.A. Lichatowich. 1991. Pacific salmon at the crossroads: Stocks at risk from California, Oregon, Idaho, and Washington. Fisheries 16(2):4-21.

Nettleton, J.A. 1990. Comparing nutrients in wild and farmed fish. Aquaculture Magazine 16(1):34-41.

Newkirk, G.F., and L.E. Haley. 1983. Selection for growth rate in the European oyster, *Ostrea edulis*: Response of second generation groups. Aquaculture 33: 149-155.

Niemi, M., and I. Taipalinen. 1982. Faecal indicator bacteria at fish farms. Hydrobio. 86:171-175.

North Atlantic Salmon Conservation Organization (NASCO). 1990. Report on the Norwegian Meeting, Loen, Norway, May.

Nose, T. 1985. Recent advances in aquaculture in Japan. Geojournal 10(3):261-276.

Officer, C.B., T.J. Smayda, and R. Mann. 1982. Benthic filter feeding: A natural eutrophication control. Mar. Ecol. Prog. Ser. 9: 203-210.

Olson, W. M. 1987. Seaweed Cultivation in Minamikayabe, Hokkaido, Japan: Potential for Similar Mariculture in Southeastern Alaska. Marine Advisory Bulletin 27. 23 pp.

Organization for Economic Co-operation and Development Fisheries Committee (OECD). 1989. Aquaculture Developing a New Industry. Fisheries Committee of the Organisation for Economic Co-operation and Development. Paris, France: OECD Publications. 126 pp.

Palmer, J.E., ed. 1989. The application of artificial intelligence and knowledge-based systems techniques to fisheries and aquaculture workshop report. Virginia Sea Grant Publication 89-03.

Palva, T.K., H. Lehvaeslaiho, and E.T. Palva. 1989. Identification of anadromous and non-anadromous salmon stocks in Finland by mitochondrial DNA analysis. Aquaculture 81(3/4):237-244.

Paramatrix, Inc. 1990. Final programmatic environmental impact statement: Fish culture in floating net-pens. Prepared for the Washington Department of Fisheries, Olympia, Wash.

Parker, N.C. 1989. History, status, and future of aquaculture in the limited status. Reviews in Aquatic Science 1(1):97-109.

Parsons, J.E., and G.H. Thorgaard. 1984. Induced androgenesis in rainbow trout. J. Exp. Zool. 231:407-412.

Parsons, J.E., and G.H. Thorgaard. 1985. Production of androgenic diploid rainbow trout. J. Hered. 76:177-181.

Pell Library. 1987. Mariculture. The National Sea Grant Depository Literature Search.

Phillips, M.J., M.C.M. Beveridge, and J.F. Muir. 1985. Waste Output and Environmental Effects of Rainbow Trout Cage Culture. International Council for the Exploration of the Sea, Vol. 21. 16 pp.

Piedrahita, R.H. 1991. Modeling water quality in aquaculture ecosystems. In Aquaculture and Water Quality, D.E. Brune and J.R. Tomass, eds. World Aquaculture Society, Baton Rouge, La.

Pillay, T.V.R. 1976. The state of aquaculture 1976. Pp. 1-10 in Advances in Aquaculture, T.V.R. Pillay and W.A. Dill, eds. Farnham, Surrey, England: Fishing News Books Ltd.

Pollnac, R.B., and J.J. Poggie. 1991. Introduction. Pp. 1-18 in Small-Scale Fishery

Development: Sociocultural Perspectives, Poggie, J. and R. Pollnac, eds. International Center for Marine Research Development, University of Rhode Island, Kingston, R.I.

Porter, C., M.D. Krom, M. Robbins, L. Brickell, and A. Davidson. 1987. Ammonia excretion and total N budget for gilthead seabream *(Sparus aurata)* and its effect on water quality conditions. Aquaculture 66: 287-297.

President's Council of Economic Advisors. 1989. Economic Report. In The Economics Report of the President. Washington, D.C.: U.S. Government Printing Office.

Price Waterhouse Management Consultants. 1990. Long term production outlook for the Canadian aquaculture industry (1990 edition) an overview. Report prepared for Department of Fisheries and Oceans, Ottawa, Ontario.

Pruder, G.D. 1991. Shrimp culture in North America and the Caribbean: Hawaii 1988. Pp. 58-69 in Shrimp Culture in North America and the Caribbean, P.A. Sandifer, ed. World Aquaculture Society, Baton Rouge, La.

Pruder, G.D. 1991. Status of shrimp farming in Texas. Pp. 36-57 in Shrimp Culture in North America and the Caribbean, P.A. Sandifer, ed. World Aquaculture Society, Baton Rouge, La.

Purdom, C.E. 1972. Induced polyploidy in plaice *(Pleurophectes platessa)* and its hybrids with the flounder *(Platichthys flesus)*. Heredity 29:11-24.

Purdom, C.E. 1983. Genetic engineering by the manipulation of chromosomes. Aquaculture 33:287-300.

Purdom, C.E. 1987. Methodology on selection and intraspecific hybridization in shellfish—A critical review. Pp. 285-292 in Selection, Hybridization, and Genetic Engineering in Aquaculture. Vol. 1., K. Tiews, ed. Berlin: Heenemann Verlagsgellschaft mbh.

Putman, J. 1989. Food Consumption, Prices, and Expenditures 1966–1987. U.S. Department of Agriculture, Economic Research Service, Statistical Bulletin No. 773. Washington, D.C.

Putnam, J.J., and J.E. Allshouse. 1991. Food Consumption, Prices, and Expenditures 1968–1989. Statistical Bulletin No. 825. U.S. Department of Agriculture, Economic Research Service, Washington, D.C.

Raible, R.W. 1979. Study of Cumulative Growth Inhibiting Factors in Recycled Water for Catfish Cultivation. Arkansas Water Resources Research Center, National Technical Information Service, PB 297-117. 44 pp.

Ratafia, M., and T. Purinton. 1989. Emerging aquaculture markets. Aquaculture Magazine (July/August):32-46.

Reeb, C.A., and J.C. Avise. 1990. A genetic discontinuity in a continuously distributed species: Mitochondrial DNA in the American oyster, *Crassostrea virginica.* Genetics 124(2):397-406.

Reed, B. 1989. Evaluation of a recirculating raceway system for the intensive culture of the penaeid shrimp *Penaeus vannamei* Boone. M.S. thesis, Department of Biology, Corpus Christi State University, Tex.

Refstie, T. 1987. Selective breeding and intraspecific hybridization of cold water finfish. Pp. 293-302 in Selection, Hybridization, and Genetic Engineering in Aquaculture, Vol. 1, K. Tiews, ed. Berlin: Keenemann Verlagsgellschaft mbh.

Refstie, T. 1990. Application of breeding schemes. Aquaculture 85:163-169.

Rensel, J.E., R.A. Horner, and J.R. Postel. 1989. Phytoplankton blooms in Puget Sound, Washington, and their effect on salmon aquaculture. Northwest Environmental Journal 22 pp.

Rhodes, R.J. 1987. Status of world aquaculture. Aquaculture Magazine 17th Annual Buyers Guide.

Rhodes, R.J., and D. Hollin. 1990. Financial analysis of commercial red drum aquaculture enterprise. Pp. 189-208 in Red Drum Aquaculture, G.W. Chamberlain et al., eds. Texas A&M University, College Station, Tex.

Richards, G.P. 1988. Microbial purification of shellfish: A review of depuration and relaying. Journal of Food Protection 51(3):218-251.

Ricker, W.E. 1969. Food from the sea. In Resources and Man, P. Cloud, ed. Chicago: Freemand and Company. 290 pp.

Rogers, G.L., and S.L. Klemetson. 1985. Ammonia removal in selected aquaculture water reuse biofilters. Aquacultural Engineering 4:135-154.

Rokkones, E., P. Alestron, H. Skjevod, and K.M. Gautvik. 1989. Microinjection and expression of a mouse metallothionein human growth hormone fusion gene in fertilized salmonid eggs. J. Comp. Physiol. B. 158:751-758.

Roland, W.G., and J.R. Brown. 1990. Production model for suspended culture of the Pacific oyster, *Crassostrea gigas*. Aquaculture 87(1):35-52.

Rosenberry, R. 1990. Shrimp farming in the Western Hemisphere. Presented at Aquatech 90, Malaysia, June.

Rosenberry, R. 1991. World shrimp farming. Aquaculture Magazine(September/October):60-64.

Rosenfield, A., and F.G. Kern. 1979. Molluscan imports and the potential for introduction of disease organisms. Pp. 165-189 in Exotic Species in Mariculture, R. Mann, ed. Cambridge, Mass.: Massachusetts Institute of Technology Press.

Rosenthal, H. 1980. Implications of transplantations to aquaculture and ecosystems. Marine Fisheries Review 42(5):1-14.

Rosenthal, H. 1985. Constraints and perspectives in aquaculture development. GeoJournal 10(3):305-324.

Rosenthal, H. 1985. Recent Observations on Environmental Effects of Large-Scale Net Cage Culture in Japanese Coastal Waters. International Council for the Exploration of the Sea. Vol. 34. 21 pp.

Rosenthal, H., D. Weston, R. Gowen, and E. Black. 1988. Report of the ad hoc Study Group on "Environmental Impact of Mariculture." International Council for the Exploration of the Sea, Cooperative Research Report No. 154. 83 pp.

Rosentreater, N. 1977. Characteristics of hatchery fish: Angling, biology, and genetics. Pp. 79-84 in Columbia River Salmon and Steelhead, E. Schwiebert, ed. Proceedings of Symposium, Vancouver, Wash. March.

Royce, W. 1989. A history of marine fishery management. Aquatic Sci. 1:27-44.

Rubino, M.C., and R.W. Stoffle. 1990. Who will control the blue revolution? Economic and social feasibility of Caribbean crab mariculture. Human Organization 49(4):386-394.

Ruttanagosrigit, W., and C.E. Boyd. 1989. Measurement of chemical oxygen demand in water of high chloride concentration. Journal of the World Aquaculture Society 20(3):170-172.

Ryther, J.H. 1969. Photosynthesis and fish production in the sea. Science 166:72-76.

Salser, B., L. Mahler, D. Lightner, J. Ure, D. Danald, C. Brand, N. Stamp, D. Moore, and B. Colvin. 1978. Controlled environment aquaculture of penaeids. In Drugs and Food From the Sea, Myth or Reality? P.M. Kaul and C.J. Sindermann, eds. Norman, Okla.: University of Oklahoma Press.

Sanbonsuga, Y., and M. Neushul. 1977. Cultivation and hybridization of giant kelps (Phaeophyceae). Pp. 91-96 in Proceedings of the Ninth International Seaweed Symposium, Jensen, A. and J. Stein, eds. Princeton: Science Press.

Sanbonsuga, Y., and M. Neushul. 1978. Hybridization of *Macrocystis* (Phaeophyta) with other float bearing kelps. J. Phycol. 14(2):214-224.

Sanders, J.E., J.L. Fryer, D.A. Leith, and K.D. Moore. 1972. Control of the infectious protozoan *Ceratomyka shasta* by treating hatchery water supplies. Prog. Fish-Culturist 34(1):13-17.

Sandifer, P.A. 1988. Aquaculture in the West: A perspective. Journal of the World Aquaculture Society 19(2):73-84.

Sandifer, P.A., J.S. Hopkins, A.D. Stokes, and R.A. Smiley. 1988. Experimental pond grow-out of the red drum, *Sciaenops ocellatus*, in South Carolina. Journal of the World Aquaculture Society 19(1):62A (Abstract).

Sandifer, P.A. 1991. Species with aquaculture potential for the Caribbean. Pp. 30-60 in Status and Potential of Aqualture in the Caribbean, J.A. Hargreaves and D.E. Alston, eds. World Aquaculture Society. 274 pp. Baton Rouge, La.

Santulli, A., E. Puccia, and V. D'Amelio. 1990. Preliminary study on the effect of short-term carnitive treatment on nucleic acids and protein metabolism in sea bass (*Dicentrarchus labrax* L.) fry. Aquaculture 87(1):85-90.

Sattaur, O. 1989. The threat of the well-bred salmon. New Scientist (April):54-58.

Saunders, R.L. 1988. Algal catastrophe in Norway. World Aquaculture 19(3):11-12.

Sawyer, E.S., P.J. Sawyer, and J.M. Lindbergh. 1990. Sea ranching of pink (*Oncorhynchus gorbuscha*) and chum (*O. keta*) salmon in the western Atlantic. Aquaculture 87:299-310.

Schiewe, M.H., A.J. Novotny, and L.W. Harrell. 1988. Virriosis of salmonids. Pp. 323-327 in Disease Diagnosis and Control in North American Marine Aquaculture, C.J. Sindermann and D.V. Lightner, eds. Amsterdam: Elsevier.

Schmick, R.A. 1988. The impetus to register new therapeutants for aquaculture. Prog. Fish-Culturist 50:190-196.

Sedgwick, Stephen. 1982. The Salmon Handbook. London: Andre Deutsch, Limited.

Seter, R.M. 1990. Potential within aquaculture issues of the use of genetic engineering and of the introduction of species, unpublished manuscript. College of Marine Studies, University of Delaware.

Shaklee, J.B., and C.P. Keenan. 1986. A practical laboratory guide to the techniques and methodology of electrophoresis and its application to fish fillet identification. CSIRO Marine Laboratories, Report 177. 59 pp.

Shang, Y.C. 1990. Socioeconomic constraints of marine aquaculture in Asia. World Marine Aquaculture 21(1):34-43.

Shiau, S., B. Sun Pan, S. Chen, H. Yu, and S. Lin. 1988. Successful use of soybean meal with a methionine supplement to replace fish meal in diets fed to milkfish *Chanos chanos* Forskal. Journal of the World Aquaculture Society 19(1):14-19.

Shigueno, K., and S. Itoh. 1988. Use of Mg-L-ascorbyl-2-phosphate as a vitamin C source in shrimp diets. Journal of the World Aquaculture Society 19(4):168-174.

Shultz, F.T. 1986. Developing a commercial breeding program. Aquaculture 57:65-76.

Shumway, S.E. 1989. A review of the effects of algal blooms on shellfish and aquaculture. International Council for the Exploration of the Sea Mariculture Committee Report. 1989/E:25. 49 pp.

Shumway, S.E. 1990. A review of the effects of algal blooms on shellfish and aquaculture. Journal of the World Aquaculture Society 21(2):65-104.

Sindermann, C.J. 1988. Disease problems created by introduced species. Pp. 394-98 in Disease Diagnosis and Control in North American Marine Aquaculture, 2nd ed., C.J. Sindermann and D.V. Lightner, eds. Amsterdam: Elsevier.

Smith, C.L. 1991. Patterns of wealth concentration. Human Organization 50(1):50-60.

Smith, T.I.J. 1986. Culture of North American sturgeons for fishery enhancement, Proceedings of the 15th U.S.-Japan Meeting on Aquaculture, Kyoto, Japan. NOAA Technical Report NMFS 85:19-27.

Smith, T.I.J., and W.E. Jenkins. 1991. Development of a shortnose sturgeon, *Acipenser brevironstrum*, stock enhancement program in North America. In Acipenser Sturgeon: Proceedings of the 1st International Bordeaux Symposium, Patrick Williot, ed. Bordeaux, France: CEMAGREF. 520 pp.

Smitherman, R.O., R.A. Dunham, and D. Tawe. 1983. Review of catfish breeding research 1969-1981 at Auburn University. Aquaculture 33:197-205.

Spence, J., D. Egan, and M. Evans. 1989. The present status and future prospects of salmon farming in British Columbia. Paper presented to the World Aquaculture Society Meeting, Los Angeles, February.

Stevens, R.E. 1982. The role of the Fish and Wildlife Service in aquaculture. Presented at the annual meeting of the California Aquaculture Association, Sacramento, January 13.

Stevens, R.E. 1984. Historical overview of striped bass culture and management. Pp. 1-15 in The Aquaculture of Striped Bass: A Proceedings, Joseph P. McCraren, ed. University of Maryland, College Park, Md. Pub. No. UM-SG-MAO-84-01.

Stickney, R. 1988. The culture of macroscopic algae. World Aquaculture 19(3): 54-58.

Stokes, R.C. 1990. Economics of salmon farming. Appendix D of final programmatic EIS, fish culture in floating net pens. Washington Department of Fisheries. Seattle.

Stolpe, N.E. Conference Proceedings from the New Jersey Aquaculture Conference. 50 pp.

Strasdine, G.A., and J.R. McBride. 1979. Serum antibiotic levels in adult sockeye salmon as a function of route of administration. J. Fish. Biol. 15:135-140.

Szyper, J., R. Bourke, and L.D. Conquest. 1984. Growth of juvenile dolphin fish, *Coryphaena hippurus*, on test diets differing in fresh and prepared components. Journal of the World Mariculture Society 15:219-221.

Talley, K. 1989. Seafood Trends Newsletter (November 13).

Taylor, M.M. 1989. Controlled purification: A policy option in the management of Washington State commercial shellfish resources. M.S. thesis. University of Washington, Seattle.

Thompson, A.G. 1990. The danger of exotic species. World Aquaculture 21(3):25-32.

Thorgaard, G.H. 1986. Ploidy manipulation and performance. Aquaculture 57:57-64.

Tibbs, J.F., R.A. Elston, R.W. Dickey, and A.M. Guarino. 1988. Studies on the accumulation of antibiotics in shellfish. Northwest Environmental Journal 5(1).

Tiddens, A.A. 1990. Aquaculture in America: The Role of Science, Government, and the Entrepreneur. Boulder Colo.: Westview Press.

Tiedje, J.M., R.K. Colwell, Y.L. Grossman, R.E. Hodson, R.E. Lenski, R.N. Mack, and P.J. Regal. 1989. The planned introduction of genetically engineered organisms: Ecological considerations and recommendations. Ecology 70: 298-315.

Tilseth, S. 1990. New marine fish species for cold-water farming. Aquaculture 85:235-245.

Toranzo, A.E., P. Combarro, M.L. Lemos, and J.L. Barja. 1984. Plasmid coding for transferable drug resistance in bacteria isolated from cultured rainbow trout. Applied and Environmental Microbiology 48:872-877.

Trimble, W.C. 1980. Production trials for monoculture and polyculture of white shrimp *(Penaeus vannamei)* with Florida pompano *(Trachinotus carolinus)* in Alabama, 1978-1979. Proc. World Mariculture Society 11:44-59.

Tseng, C.K. 1981. Commercial Cultivation. Pp. 680-725 in Biology of Seaweeds, C. Lobban and M. Wyne, eds. Berkeley: University of California Press.

Tucker, C.S., and C.E. Boyd. 1985. Water quality. Pp. 135-227 in Channel Catfish Culture, C.S. Tucker, ed. Amsterdam: Elsevier.

Tucker, C.S., and S.W. Lloyd. 1985. Water Quality in Streams and Channel Catfish *(Ictalurus puntatus)* Ponds in West-Central Mississippi. Mississippi Agriculture and Forestry Experimental Station, Technical Bulletin 129, 8 pp.

Tucker, C.S., C.E. Boyd, and E.W. McCoy. 1979. Effects of feeding rate on water quality, production of channel catfish, and economic returns. Trans. Amer. Fish. Soc. 108:389-396.

Urner-Barry. Various issues. Seafood Price: Current. Tom's River, New Jersey.

U.S. Bureau of the Census. 1988. Unpublished data. Microfiche Series, Foreign Trade Division, EM575 and IM-175.

U.S. Congress. 1977. Food and Agriculture Act of 1977. Statues at Large 91, p. 1021.

U.S. Department of Agriculture. 1988. Aquaculture Genetics and Breeding, National Research Priorities, Vol. II. Office of Aquaculture. Washington, D.C. 61 pp.

U.S. Department of Agriculture. 1988. Aquaculture: Situation and Outlook Report. Washington, D.C. 39 pp.

U.S. Department of Agriculture. 1990. Outlook for U.S. Agricultural Exports. Foreign Agricultural Service, Economic Research Service, Washington, D.C.

U.S. Department of Agriculture/U.S. Department of the Interior. 1990. Final report of the USDA-USDI protective statutes workgroup. December 1990 (unpublished report).

U.S. Department of Commerce. 1988. Aquaculture and Capture Fisheries: Impacts in U.S. Seafood Markets. National Organization of Aquaculture Associates, National Mariculture and Fisheries Society, Washington, D.C.

U.S. Department of Commerce. 1990. Fisheries of the United States. National Organization of Aquaculture Associates, National Mariculture and Fisheries Society, Washington, D.C.

van de Meer, J.P., and M.V. Patwary. 1983. Genetic modification of *Gracilaria tikvahiae* (Rhodophycease). The production and evaluation of polyploids. Aquaculture 33:311-316.

Van Engel, W.A., W.A. Dillon, D. Zwerner, and D. Eldridge. 1966. *Loxothylacus panopei* (Cirripedia, Sacculinidae) an introduced parasite on a xanthid crab in Chesapeake Bay, U.S.A. Crustaceana 10:110-12.

Van Olst, J.C., and J.M. Carlberg. 1990. Commercial culture of hybrid striped bass. Aquaculture Magazine 16(1):49-59.

van Toerer, W., and K.T. Mackay. 1981. A modular recirculation hatchery and rearing system for salmonids utilizing ecological design principles. Pp. 403-413 in Aquaculture in Heated Effluents and Recirculation Systems, Vols. 16-17, K. Tiews, ed. Schriffen der Bundesforschgsanstalt fuer Fishererei.

Vermeij, G.J. 1991. When biotas meet. Understanding biotic interchange. Science 253:1099-1103.

Wada, K.T. 1986. Genetic selection for shell traits in the Japanese pearl oyster *Pinctada fucata martensii*. Aquaculture 57:171-176.

Walton, J.R. 1988. Antibiotic resistance: An overview. Veterinary Record 122:247-251.

Wang, J.K. 1988. Shared resources aquatic production systems. American Society of Agricultural Engineers, ASAW Paper No. 88-5001. St. Joseph, Michigan.

Wang, J.K. 1990. Managing shrimp pond water to reduce discharge problems. Aquacultural Engineering 9:61-73.

Waples, R.S., G.A. Winans, F.M. Utter, and C. Mahnken. 1990. Genetic approaches to the management of Pacific salmon. Fisheries 15(5):19-25.

Water Farming Journal. 1991. FDA adopts tough new policy on use of drugs in aquaculture. (September 28):4-6, 27.

Webb, A.J., M. Lopez, and R. Penn. Estimates of Producer and Consumer Subsidy Equivalents: Government Intervention in Agriculture, 1982-89. USDA Educational Research Service. Statistical Bulletin No. 803. Washington, D.C. 358.

Wedemeyer, G.A., N.C. Nelson, and C.A. Smith. 1978. Survival of salmonid viruses infectious hematopoietic necrosis (HNV) and infectious pancreatic necrosis (IPNV) in ozonated, chlorinated, and untreated waters. J. Fish. Res. Bd. Can. 35:875-879.

Weeks, P. 1990. Marine aquaculture development: An anthropological perspective. World Marine Aquaculture 21(3):69-74.

Wenk, E., Jr. 1972. The Politics of the Ocean. Seattle: University of Washington Press,

Westley, R.E., N.A. Rickard, C.L. Goodwin, and A.L. Scholz. 1990. Enhancement of molluscan shellfish in Washington State. Pp. 49-52 in Marine Farming and Enhancement. Proceedings of the 15th U.S.-Japan Meeting on Aquaculture, A.K. Sparks, ed. Kyoto, Japan, National Organization of Aquacultural Associates Technical Report NMFS 85.

Weston, D.P. 1986. The environmental effects of floating mariculture in Puget Sound. School of Oceanography, College of Ocean and Fishery Science, University of Washington, Seattle. 148 pp.

Weston, D.P. 1991. An environmental evaluation of finfish net-cage culture in Chesapeake Bay. Report of the Center for Environmental and Estuarine Studies, Horn Point Environmental Laboratory, Cambridge, Md. 78 pp.

Weston, D.P. 1991. The effects of aquaculture on indigenous biota. Pp. 534-67 in Aquaculture and Water Quality, D.E. Brune and J.R. Tomasso, ed. Baton Rouge, La.: The World Aquaculture Society.

Weston, D.P., and R.J. Gowen. 1988. Assessment and prediction of the effects of salmon net-pen farming on the benthic environment. Report to Washington Department of Fisheries, Olympia, Wash. 62 pp.

Wheaton. 1977. Aquacultural Engineering, Wiley-Interscience Publication. New York: John Wiley and Sons.

Wheaton, F.W. 1985. Aquacultural Engineering. Malabar, Fla.: Robert E. Krieger.

Whetstone, J.M., E.J. Olmi III, and P.A. Sandifer. 1988. Management of existing saltmarsh impoundments in South Carolina for shrimp aquaculture and its implications. Pp. 327-338 in The Ecology and Management of Wetlands, D.D. Hook et al., eds. Portland, Ore.: Timber Press.

Whitaker, C., and R.B. Fridley. 1987. A simulation model for evaluating predation control alternatives in salmon ocean-ranching. Automation and Data Processing in Aquaculture. IFAC Proceedings Series 1989(9):125-131.

Whitehurst, D.K., and R.E. Stevens. 1990. History and overview of striped bass culture and management. Pp. 1-6 in Culture and Propagation of Striped Bass and Its Hybrids, R.E. Horol, J.H. Carbo, and R.V. Minton, eds. Striped Bass Committee, Southern Division, American Fisheries Society, Bethesda, Md.

Whiteley, A.H., and A. Johnstone. 1990. Additives to the environment of net-pen reared fish. Proc. Pacific Marine Fisheries Commission 42nd Annual Meeting, Seattle, October 16-18, 1989.

Wiesmann, D., H. Scheid, and E. Pfeffer. 1988. Water pollution with phosphorus of dietary origin by intensively fed rainbow trout *(Salmo gairdneri* Rich.). Aquaculture 69(3/4):263-270.

Wildsmith, B.H. 1982. Aquaculture: The Legal Framework. Toronto: Edmond Montgomery Publications Ltd. 305 pp.

William, M.L., and R.C. Heidinger. 1981. Tank Culture of Striped Bass. Illinois Striped Bass Project, Fisheries Research Laboratory, Southern Illinois University. Report IDC F-26-R. 115 pp.

Williams, R.R., and D.V. Lightner. 1988. Regulatory status of therapeutics for penaeid shrimp culture in the United States. Journal of the World Aquaculture Society 19(4):188-196.

Wilson, J., and D. Fleming. 1989. Economics of the Maine mussel industry. World Aquaculture 20(4):49-55.

Wolniakowski, K., M. Stephenson, and G. Ishikowa. 1987. Tributyltin concentrations and oyster deformations in Coos Bay, Oregon. Pp. 1438-1442 in Oceans '87 Proceedings, Vol. 4, International Organotin Symposium.

Wolters, W.R., G.S. Libey, and C.L. Chrisman. 1981. Induction of triploidy in channel catfish. Trans. Am. Fish. Soc. 110:310-312.

Wolters, W.R., G.S. Libey, and C.L. Chrisman. 1982. Effect of triploidy on growth and gonad development of channel catfish. Trans. Amer. Fish. Soc. 111:102-105.

Wray, T. 1990. World's biggest trout farmer. Fish Farming International 17(1): 18-23.

Wright, K. 1990. Bad news bacteria. Science 249:22-24.

Wyban, J.A., and E. Antill, eds. 1989. Instrumentation in Aquaculture. Proceedings of a Special Session at the World Aquaculture Society 1989 Annual Meeting, Oceanic Institute, Hawaii. 101 pp.

Wyban, J.A., and J.N. Sweeney. 1991. Intensive Shrimp Production Technology; the Oceanic Institute Shrimp Manual. The Oceanic Institute, Hawaii. 158 pp.

Yamada, S., Y. Tanaka, M. Sameshima, and Y. Ito. 1989. Pigmentation of prawn *(Penaeus japonicus)* with carotenoids. Effect of dietary astaxanthin, β-carotene, and canthaxanthin on pigmentation. Aquaculture 87(3/4):323-330.

Yamazaki, F. 1983. Sex control and manipulation in fish. Aquaculture 33:329-354.

Ziemann, D., G.D. Pruder, and J.K. Wang. 1990. Honolulu, Hawaii: Aquaculture Effluent Discharge Program Year 1 Final Report. Center for Tropical and Subtropical Aquaculture, 212 pp.

Appendix A

Review of World Aquaculture

MAJOR WORLD AQUACULTURE PRODUCTS

Finfish Culture

Table A-1 provides an overview of world aquaculture production in 1987 by region and type of seafood. Marine and freshwater species are combined in the available statistics on world production. Historically, world aquaculture has been dominated by the pond culture of freshwater finfish, particularly the various species of carp (common, Chinese, Indian) grown throughout Europe and Asia. Together, all varieties of carp still account for approximately 4 million metric tons (mmt), or more than one-quarter of the world's annual production of finfish cultured in fresh water. More recently, other species of finfish have contributed significantly to freshwater aquaculture, such as tilapia (0.3 mmt), rainbow trout (0.2 mmt), channel catfish (0.2 mmt), and eel (0.1 mmt). These, and minor contributions from crayfish culture (0.03 mmt) and the giant freshwater prawn *Macrobrachium* (0.03 mmt), bring the total production of freshwater organisms to nearly 5 mmt, or about one-third of the world's total.

The cultivation of marine finfish has lagged far behind that of freshwater species. The oldest such practices are the growing of milkfish in the Philippines, Indonesia, and other tropical Asian countries (0.3 mmt) and of the yellowtail (amberjack) in Japan (0.2 mmt). Milkfish are grown in shallow estuarine ponds; yellowtail, in net cages. In both cases, rearing technology is relatively crude. Neither species can be matured or spawned routinely in captivity, so the industries are based on the collection of fry (juveniles) from the wild. Both species are still fed primarily natural food.

TABLE A-1 World Aquaculture Production, 1987 (metric tons)

Region	Finfish	Crustaceans	Mollusks	Seaweed	Other	Total	Percentage of World Total
Africa, north and northeast	51,397	2	286		50	51,685	0.39
Africa, south of Sahara	10,461	77	229			10,817	0.08
North America	266,672	44,480	138,841			449,993	3.41
Central America	9,485	6,564	50,719			66,768	0.50
South America	21,674	79,759	2,456	9,178		113,067	0.86
Caribbean	17,725	800	1,503	210	31	20,269	0.15
Europe	399,037	3,285	645,271			1,047,593	7.93
USSR	288,970		159	3,459		292,588	2.22
Near East	23,816	18				23,834	0.18
Oceania	2,730	174	27,185	1,710	140	31,939	0.24
East Asia	4,421,638	326,436	1,720,092	3,064,916	27,461	9,561,353	72.39
West Asia	1,279,836	113,311	84,843	60,000	20	1,538,010	11.64
Total	6,793,441	574,906	2,672,394	3,139,473	27,702	13,207,916	
Percentage of world total	51.4	4.4	20.2	23.8	0.2	100	

SOURCE: Food and Agriculture Organization (1989).

A more recent development in marine finfish culture is the growth of salmonids in net pens of cages in protected coastal waters. For many years, salmon have been hatchery spawned and reared to smolt size, at which stage they are physiologically adapted for introduction to salt water; they are then released to the ocean to enhance natural stocks. Beginning about 1970, attempts were made to rear Pacific salmon smolt in cages in the Puget Sound area of Washington. The initial product, a 150- to 250-gram (g) "pan-sized" salmon that could be reared in one growing season, did not prove to be very successful in the marketplace, and the practice gradually dwindled.

A decade later, Norwegians initiated the net-cage culture of Atlantic salmon in their large fjord systems, this time holding the fish for two growing seasons, and feeding them a carefully formulated pelleted diet until they reached a size of 4–5 kilograms (kg). The practice proved highly successful, a single 12 × 12 meter cage producing as much as 5–10 tons of salmon over an 18-month grow-out period. Atlantic salmon cage culture has now spread to the United Kingdom, France, Spain, both coasts of the United States and Canada, Chile, Australia, and New Zealand; in 1988, nearly 0.2 mmt were produced worldwide.

Several other marine finfish are currently grown successfully in smaller quantities in various parts of the world. These include gilthead sea bream, sea bass, and turbot in Europe; aiyu, flounder, puffer fish, red and black sea bream, and several other species in Japan; and the estuarine grouper in Malaysia, Singapore, and Hong Kong. Together, annual production of the several marine finfish now in culture probably approaches 1 mmt, only about 20 percent of freshwater finfish culture.

Crustacean Culture

The culture of crustaceans is almost entirely restricted to two groups of shrimp or prawns: the giant freshwater prawn *Macrobrachium* and several species of marine shrimp of the genus *Penaeus*.

After development of the technology for rearing *Macrobrachium* in Malaysia in the early 1960s, there was much interest in growing the species throughout the world's tropics. Interest has flagged during the past decade, due more to marketing than to technical problems and to the inability of the product to compete with marine species. *Macrobrachium* spp. are still grown successfully in a number of small scattered operations, with production totaling approximately 0.02 mmt.

Marine (penaeid) shrimp culture, on the other hand, is one of the fastest growing and economically most successful forms of aquaculture in practice today. The technology for hatchery spawning of gravid (fertile) female penaeids and controlled rearing of their larvae in captivity, were first developed more than 50 years ago in Japan and rapidly spread throughout Southeast Asia. When it was found that the postlarvae could be grown out quickly and easily to a marketable adult in shallow estuarine ponds, shrimp culture spread to the Philippines and Indonesia.

By the late 1970s, the shrimp farming industry had spread to the Western Hemisphere, where it was initially centered in the extensive estuarine system of Ecuador's Guayas River and Gulf of Guayaquil. Subsequently the industry moved into virtually all of the tropical maritime countries of South and Central America, Mexico, and the southern parts of the United States. However, Ecuador remains the major producer in the Western Hemisphere.

As in the case of shrimp farming in Asia, the Latin American industry is based on the collection of postlarvae from the wild, supplemented as necessary by hatchery production of young from wild-caught gravid females. Dependence on wild postlarvae or gravid females restricts the location of shrimp farms to coastal regions, where natural populations occur in abundance. Even in those locations, supplies may be erratic and undependable, and they may disappear entirely with the onset of unfavorable climatic phenomena such as the South American El Niño.

Shrimp aquaculture is the production of shrimp involving control of one or more phases of their biological cycle or control of the environment in which they develop. Management systems may be extensive, such as large seminatural or natural marsh impoundments or rice fields (low stocking rates and little or no feeding and water exchange); semi-intensive, such as large drainable ponds (medium stocking, feeding, and water exchange); or intensive, such as small, highly controllable ponds (high stocking rates, water circulation and exchange, and nutritionally complete diets). Indoor raceways would exhibit the highest degree of technology, with control of nutrition and environmental requirements for year-round growth. However, to date no indoor raceways have proved economically viable for commercial production of shrimp.

Worldwide, the significance of shrimp aquaculture has increased dramatically over the past decade. In 1980, only about 2 percent of the world's shrimp supply was produced by aquaculture, whereas by 1990 farmers were supplying 25 percent of the market (Table A-2) (Rosenberry, 1991a). Most

TABLE A-2 World Shrimp Production, 1991 (heads-on)

Source	Amount (mmt)	Percentage of World Production
Fisheries (4.327 billion lb)	1.4967	75
Aquaculture (1.393 billion lb)	0.633	25

SOURCE: Rosenberry (1991b).

TABLE A-3 Summary of World Shrimp Production by Hemisphere, 1991

	Percentage of World Production	Heads-on Production (metric tons)	Area in Production (hectares)	Yield (kg/hectare)	Number of Hatcheries	Number of Farms
Eastern Hemisphere	81	556,500 (1.224 billion lb)	819,500 (2,024,165 acres)	679 (605 lb/acre)	4,501	34,840
Western Hemisphere	19	133,600 (0.294 billion lb)	129,450 (319,741 acres)	752 (670 lb/acre)	207	2,055
Total	100	690,100 (1.518 billion lb)	1,022,950 (2,526,686 acres)	630 (561 lb/acre)	4,708	36,845

SOURCE: Rosenberry (1991b).

(81 percent) of this production is concentrated in the Eastern Hemisphere (Southeast Asia) (Table A-3). A fairly recent and dramatic entry to shrimp farming has been mainland China, which in just a decade or so has become the world's leading producer of aquacultured shrimp (Table A-4)(Rosenberry, 1991a,b).

Western Hemisphere production is led by Ecuador (Aiken, 1990; Rosenberry, 1991a,b), which contributes 75 percent of the region's farmed production, whereas the United States produces only about 1 percent (Table A-5). China

TABLE A-4 Eastern Hemisphere Shrimp Production, 1991

Country	Production (%)	Area in Production (acres)	Yield (lb/acre)	Number of Hatcheries	Number of Farms
China	26.1	345,800	923	1,000	2,000
Indonesia	25.2	494,000	623	250	20,000
Thailand	19.7	197,600	1,255	2,000	3,000
India	6.3	160,550	479	16	2,500
Philippines	5.4	123,500	534	250	3,000
Vietnam	5.4	395,200	167	120	1,000
Taiwan	5.4	19,760	3,340	800	2,000
Bangladesh	4.5	247,000	223	0	1,000
Other	1.4	39,520	445	25	175
Japan	0.6	1,235	6,235	40	165
Total	100	2,024,165	602	4,501	34,840

SOURCE: Rosenberry (1991b).

TABLE A-5 Western Hemisphere Shrimp Production, 1991

Country	Production (%)	Area in Production (acres)	Yield (lb/acre)	Number of Hatcheries	Number of Farms
Ecuador	74.9	358,150	615	150	1,700
Colombia	6.7	9,880	2,004	20	30
Mexico	3.7	12,350	891	6	100
Honduras	3.4	17,290	573	2	25
Panama	3.0	9,880	891	6	40
Peru	2.6	9,880	779	3	60
United States	1.2	1,112	3,167	3	25
Other	4.5	11,856	1,113	17	75
Total	100	430,398	683	207	2,055

SOURCE: Rosenberry (1991b).

has constructed some 1,000 hatcheries to produce juvenile shrimp for stocking in aquaculture ponds; Ecuador has 150 hatcheries; the United States has only 3 (Tables A-4 and A-5). Overall, Asia has approximately 4,500 hatcheries, whereas the Western Hemisphere has a little more than 200 (Table A-3).

Mollusk Culture

Bivalve mollusks (i.e., oysters, clams, mussels, scallops) are sessile, grow without confinement, and feed on natural food organisms (i.e., unicellular algae suspended in the water). Their cultivation on privately owned or leased bottom is therefore simple and inexpensive, and differs little from capture fishing on public grounds. In either case, aquaculture is involved if and when natural stocks become depleted and must be enhanced by reseeding the bottom.

Cultivation of seed in hatcheries is a well-developed technology for most important commercial species of mollusks that was initiated in the United States in the 1920s and is now widely practiced around the world. However, hatchery production of seed is costly; it may often be avoided by collecting natural seed or enhancing the natural set of seed by placing seed or spat collectors at strategic times and locations in the growing area. Shellfish larvae are, of course, most abundant where there are large populations of adult animals, as at major aquaculture sites, most of which may consequently collect their own seed without recourse to hatcheries.

The Japanese discovered many years ago that oysters could be grown much more quickly and abundantly in a three-dimensional mode, from surface to bottom, on ropes or wires suspended from rafts. The shellfish thereby have access to a much greater supply of food; they are protected from sedimentation and benthic predators; and vastly more animals may be grown per unit area than by traditional bottom culture methods. Raft culture of oysters, mussels, and scallops is now common practice in those countries that are the leading producers of bivalve mollusks (i.e., Japan, China, Taiwan, North and South Korea, Spain). In most if not all such operations, natural seed is collected at or near the culture site.

Of the 3 mmt of mollusks cultured in 1988, 67 percent were grown in East Asia and another 20 percent in Europe (Table A-1). The Pacific cupped oyster *(Crassostrea gigas)*, grown most abundantly in Japan but now successfully introduced around the world, is the leading cultured bivalve species (0.8 mmt). The several species of mussels now grown in many different countries together yielded 1 mmt; various clam species accounted for another 0.4 mmt; and the Japanese scallop, first grown in that country only about five years ago, had already contributed 0.3 mmt by 1988 (FAO, 1990).

If the Japanese experience is typical, scallops, which grow extremely fast and command a high market price, may overtake other bivalve species as a

Stake culture of oysters in Japan.

favored culture product. Scallop farms have now been started up in Peru, Chile, Canada, the United States, and China. Bivalve mollusk farming is, by far, the most successful form of marine animal culture; it is more than twice as productive as finfish and crustacean culture combined. Despite the strong emphasis and publicity given to penaeid shrimp culture in recent years, mollusk farming has been advancing much more rapidly, and the value of the product is, in most cases, equally attractive.

Seaweed Culture

Different species of seaweed have long been regarded as both luxury and staple foods or food supplements in many Asian countries. The red alga *Porphyra,* grown and marketed as nori in Japan, is among the most costly of seafood. The kelp *Laminaria,* also grown in Japan, was formerly exported in quantity to China, whose inhabitants are susceptible to the glandular disease goiter, caused by an iodine deficiency, a condition remedied by a seaweed dietary supplement. Today, China grows more than 1 mmt (dry weight) of *Laminaria* annually and now exports part of the crop to Japan (Tseng, 1981).

In addition to their direct use as human food, many seaweeds contain the polysaccharides agar, algenic acid, or carrageenan, which, when extracted from the plants, have widespread commercial value as emulsifying or suspending agents in the food, drug, and cosmetic industries. The dwindling supply of wild stock of these seaweeds has led to their cultivation in several countries. About one-half of the Chinese crop of *Laminaria* is used for extraction of algenic acid; the other half is consumed as food. The red seaweed *Gracilaria* is grown in Taiwan and Chile as a source of agar; another red algae, *Eucheuma,* is cultured in the Philippines for its carrageenan content.

A total of 4 mmt of seaweed was grown worldwide in 1988, 90 percent in Asia. This is the largest group, by weight but not by value, of cultivated marine organisms, representing 25 percent of the total and about 42 percent of the marine component.

It should be pointed out that seaweeds are sometimes not included in aquaculture statistics (an omission that may lead to considerable confusion when comparing data). If seaweed is omitted from consideration, total world aquaculture production for 1988 was 11 mmt., of which only about 5 mmt (43 percent) were from marine aquaculture.

TABLE A-6 Value of Aquaculture Production by Leading Countries, 1984–1987 (hundred U.S. dollars)

	1984	1985	1986	1987
China	4,059,465	4,788,214	5,440,725	6,078,454
Japan	2,263,753	2,279,309	3,439,373	3,895,790
Taiwan PC	607,576	631,672	818,655	1,110,282
India	746,300	746,300	746,300	746,300
United States	500,403	429,410	484,211	563,649
Philippines	446,639	468,332	511,182	560,317
USSR	357,279	365,596	464,481	537,767
Ecuador	235,200	211,435	214,781	510,671
France	226,178	243,857	425,298	474,033
Vietnam	327,400	375,600	433,080	459,480
Korea, Republic of	253,278	266,222	327,310	438,560
Korea, Democratic People's Republic	398,700	420,700	420,700	420,700
Indonesia	264,805	351,393	375,427	385,740
Norway	133,357	177,534	233,343	314,348
Italy	140,008	140,247	185,612	238,153
Thailand	105,989	114,848	147,485	237,803
Other	—	—	—	—
Total world production	12,430,634	13,592,725	16,345,988	18,911,991

SOURCE: FAO Fisheries Circular No. 815, Rev. 1, 1989.

Economics of World Aquaculture

The monetary value of the 1988 world aquaculture crop of 14 mmt is estimated to be $22.5 billion (U.S.), an increase of 19 percent from the $18.8 billion value of the 1987 crop and more than twice that of the 1985 yield ($13.1 billion) (FAO, 1990) (Table A-6). Of the 144 countries that now report aquaculture statistics to the Food and Agriculture Organization (FAO) of the United Nations, only 60 provided information on prices and value, so total value estimates are based just on those reports. However, those 60 include most of the total production.

The value of world aquaculture has clearly increased much more rapidly than the size of the crop, probably owing mainly to world inflation. Some of the increased value may result from recent emphasis on high-priced luxury species (i.e., mollusks, shrimp, salmon), but the overall increase in production has resulted as much from low-value crops, such as seaweed, as from more expensive items.

MARINE AQUACULTURE PRACTICES AND POLICIES

Throughout the world, the most common form of marine aquaculture is carried out by collecting and growing "wild seed" in ponds, cages, or other enclosures, with the addition of fertilizer and, in some cases, food. Culture practices are primitive, labor and monetary inputs are small, and production is low. Such extensive marine aquaculture is practiced in warm parts of the world in countries that have ample available coastal waters and a traditional marine diet. Even in areas where intensive culture is practiced, the industry often depends on collections from the sea in the form of gravid females, fertilized eggs, spat, postlarvae, or juvenile animals. This practice sooner or later conflicts with fisheries resources and is not a viable alternative for the U.S. aquaculture industry.

The trend in developed countries is toward intensive culture, with breeding, rearing, and harvesting in controlled facilities using high stocking densities and formulated feeds. The ultimate objective is regular production of a high quality product at a designated time, independent of season. Examples of marine animals under intensive culture are red sea bream, yellowtail, and Japanese flounder in Japan; sea bream, sea bass, and turbot in temperate European countries; European eel and salmon in northern Europe, Japan, North America, Chile, and New Zealand; and the banana prawn in Singapore, as well as the tiger prawn *(Penaeus monodon)* throughout Southeast Asia (Juario and Benitez, 1988; Gousset, 1990). Other species under investigation for intensive culture include cod, halibut, and dolphin (Tilseth, 1990).

Bridging the gap between extensive and intensive systems, ocean ranching is best known in the production of salmon and is used when the cultured species does not stray, as is the case with oysters and abalone, or when a species can be trained or migrates naturally to return to a site, as with some salmon. Ocean ranching may be the system of choice for most mollusk and algae culture. Similar to ocean ranching is stock enhancement, the production of large numbers of young that are released to the sea and harvested by ordinary fisheries methods. The value of such stocking programs in increasing fisheries production is still under debate, but stocking is being carried out in the United States as well as in Japan and Norway.

Aquaculture goals vary from country to country but generally include the following:

- generation of needed and inexpensive protein;
- reliable production of quality products not readily available from naturally occurring sources;
- expansion of foreign trade by increasing exports or reducing imports;
- development of new industry and jobs; and
- enhancement or maintenance of fishery resources through stocking.

All but the first of these goals are important in driving aquaculture development in the United States.

ROLE OF GOVERNMENT

Government has played a pivotal role in aquaculture development in many countries that have become world leaders in marine aquaculture, including Norway, Denmark, France, Canada, and Japan. For the most part, these groups are concerned with both fisheries and marine aquaculture, and much of the aquaculture development emanates from a fisheries management perspective. Long-range planning and a strong commitment by the central government in Norway, and by central and provincial governments in Canada, have been responsible for the rapid growth of marine aquaculture in those countries. A good example is the Canadian government's support of the emerging pen-raised salmon industry in New Brunswick. Impressive growth of this industry was in part a result of $5 million in industry development funds from the New Brunswick government, the establishment of a government-funded demonstration farm, and the provision of grants and extension services (Bettencourt and Anderson, 1990).

STATUS OF MARINE AQUACULTURE BY REGION

Asia

The Asia-Pacific region is the center of development of world aquaculture and accounts for approximately 80 percent of world aquaculture production (Juario and Benitez, 1988). China is by far the largest producer, accounting for slightly more than 50 percent of the finfish (all freshwater) and more than 25 percent of all crustacean, molluscan, and seaweed world production. In China as in many other Asian countries, aquaculture technology, for the most part, is simple, utilizing natural resources and an abundant labor force to grow out products in extensive or semi-intensive systems. An exception is the application of highly advanced genetics techniques (i.e., cell culture, gene transplants) to develop new and disease-resistant strains of carp. Recently, some large commercial investments have been made in new and intensive farms for marine shrimp.

Japan is a major producer and consumer of aquaculture products, taking more than 1 million tons in 1988. Principal species for culture include red sea bream (*Pagrus major*), black sea bream (*Acanthopagrus schelegi*), yellowtail (*Seriola guinqueradiata*), Japanese flounder (*Paralichthys olivaceus*), puffer fish (*Takifugu rubripes*), Kuruma prawn (*Penaeus japonicus*), abalone (*Nordicus discus*), blood ark shell (*Scapharca broughtonii*), and edible seaweeds (*Porphyra, Undaria, Laminaria*).

Production of several species (i.e., coho salmon, rainbow trout, oyster, and laver, a seaweed) depends entirely on culture. Growing inedible animals such as pearl oysters and ornamental fish is an important aspect of aquaculture in Japan and elsewhere. Like the culture of ornamental fish in the United States, it is already a thriving industry that could eventually be expanded to include many marine fish.

Culture techniques for finfish, in general, include spawning fish in captivity either with hormone injections or with temperature and/or photoperiod control, intensive culture through the larval stage, and grow-out to marketable size in floating net cages. Production constraints are due to environmental deterioration around the farming grounds that can retard growth and cause mortality. Shrimp culture in Japan still relies heavily on wild stock, and the scarcity of gravid females greatly influences prices and brood stock production. Larval shrimp are reared in high-density, intensive systems, and fry are grown to marketable size in ponds. Production constraints include the lack of reliable egg production and the need for a practical diet to substitute for live food to rear the young. In light of these problems, it is interesting to note that the banana prawn (*Penaeus merguiensis*) in Singapore can be cultured intensively and the brood stock can be spawned in captivity, so many of the constraints on shrimp culture are removed.

Of the 19 species of shellfish reared in Japan, 12 are propagated artificially. Hatchery-bred blood ark shell and noble scallop are cultured artificially to marketable size, and abalones are usually released into the sea. Other species are grown by collecting wild spat and transplanting them to grow-out areas. It has become increasingly difficult to procure sufficient numbers of spawners from the wild. Production could be enhanced by techniques to induce maturation and to rear cultured seed to adult spawners. Japan is also culturing coho salmon in net pens, chum salmon for ocean ranching, a species of *Paralichthys* flounders, red sea bream, and abalone.

Seaweed culture is based on wild collection of spores, followed by laboratory culture of seedlings, with transfer of buds to string or nets and grow-out in coastal waters. Production constraints are weather conditions and disease. Deterioration of the water around farms occurs with long-term, high-density culture, resulting in slow growth and disease.

Constraints on aquaculture development in Asia are environmental (e.g., lack of suitable sites, pollution, exposure to natural hazards, and human factors), biotechnological (e.g., dependence on wild stocks, inadequate feed, disease outbreaks), and socioeconomic (e.g., user conflicts in coastal zones, lack of institutional support, limited demand, and market saturation). Some solutions to these problems lie in research and development activities focused on fish diets, planned production, quality control, and new markets (Mito and Fukuhura, 1988).

Northern Europe

The major contribution from northern European countries has been the development of net pen culture of salmon and highly intensive culture in raceways and tanks for salmonids and the European eel *(Anguilla anguilla)*. At present, a concerted effort is under way in the region to develop mass culture of cod (*Gadus morhua*) and Atlantic halibut (*Hippoglossus hippoglossus*). The Norwegian Fisheries Research Council and the Fish Farmers Sales Association began a national research program in 1987 with the objective of developing economically feasible methods for farming cold water species. These two species, along with the wolffish *(Anarhichas lupus)*, were found to be the most promising marine species. Brood stock cod have been domesticated, and a method for stripping captive halibut has been established. Rearing will probably be carried out in large plastic bags floating in enclosures with grow-out (to market size) in cages. Inadequate knowledge of larval nutrition at the first exogenous feeding is the main constraint in mass rearing of these cold water species (Tilseth, 1990). Expertise in controlled culture conditions and in environmental studies continues to be an important area for aquaculture development in Europe.

An excellent example of the application of technology to aquaculture is the development of modern eel farming. Denmark was a major producer of European eel, but in the late 1970s, production fell drastically while exports remained high. Denmark had to import eels to support its export industry, a situation that caused the Danish Water Quality Institute (followed by the Danish Aquaculture Institute) to develop eel farming (Gousset, 1990). Because eels require high water temperatures, energy requirements were met by growing them in recirculating systems in insulated buildings. For the effort to be profitable, rearing densities were increased so that the use of pure oxygen and water purification were required. This led to the development of advanced recirculation systems with suspended solid removal by sieving and ammonia removal by biological filters. Fish are fed with self-feeders or automatic feeders and may reach a biomass as high as 200 kg per cubic meter. These highly sophisticated systems allow farmers in northern Europe to grow eels successfully under severe climatic conditions and at the same time to greatly reduce the effluents and waste discharged into the environment (Gousset, 1990).

Net-pen culture of Atlantic salmon *(Salmo salar)* is practiced in Norway, Scotland, Ireland, and the Faroe Islands. In Norway, the leading world producer of farmed salmon, aquaculture ventures developed in the early 1970s, and industry's output has since grown exponentially from 500 metric tons in 1971 to an estimated 120,000 metric tons in 1989.

In Norway, about 20 farms closed in 1989 as a consequence of pressure from government and creditors, and an additional 50 to 70 farms were expected to follow. As a consequence of the market crisis, both Norway and Scotland (the second largest producer of farmed salmon) decreased the number of smolts stocked during that year, and production from those two regions was expected to level off for 1990-1991 (Needham, 1990). The vulnerability of the industry to unfavorable market conditions illustrates one of the major problems that faces salmon aquaculture's farmers and investors: the long life cycles (three years from egg to market-size adult for Atlantic salmon) force firms to make production decisions (such as the number of smolts to stock in any particular year) in many cases before accurate price forecasts can be made for the timing of the harvest. This leaves the industry extremely exposed to unstable marketing conditions, which are made even more volatile by the unpredictability of wild catches.

Central and Southern Europe

In recent years, considerable progress has been made in the development of mass rearing of European temperate marine fish species such as European sea bass (*Dicentrarchus labrax*), turbot (*Scophthalmus maximus*), and gilthead sea bream (*Sparus aurata*). Rapid expansion in marine fish farming has accompanied improvements in larval rearing techniques and financial support from national governments as well as the European Community. With this growth established, fish farms faced with increased competition from newcomers are searching for new markets, trying to improve growth rates, and diversifying with new species. It is expected that through improved nutritional quality of live foods and better hygiene procedures, survival rates of an increasingly diverse group of warm water finfish will improve.

Aquaculture production in the European Community (EC) reached 847,000 metric tons in 1989, worth more than 7,900 million ECUs. Finfish production is dominated by rainbow trout *(Onchorhynchus mykiss),* with production levels reaching 144,000 metric tons in 1989 and cultivated in most countries, and Atlantic salmon *(Salmo salar),* cultivated in the United Kingdom, Ireland, France, and Spain (35,000 metric tons in 1989). Other species commonly cultivated are carp *(Cyprinus carpio),* catfish (Ictaluridae), European eel *(Anguilla anguilla),* and increasingly sea bass *(Dicentrarchus labrax),* sea bream (*Sparus aurata* and *Diplodus* spp.), and turbot *(Scophthalmus maximus).* Additionally, mullet *(Mugil)* and yellowtail *(Seriola)* are cultivated on a smaller scale, along with, on an experimental scale, sturgeon (*Acipenser* spp.) and halibut *(Hippoglossus hippoglossus).*

Among the shellfish, production is dominated by mussels *(Mytilus* spp.), raised in either bottom culture (Ireland, the United Kingdom, Netherlands, Germany, and France) or rope culture (the United Kingdom, Ireland, Spain, France, and Italy), and oysters *(Ostrea edulis* and *Crassostrea gigas).* Clam culture *(Ruditapes phillippinarum* and *Tapes semidecussata)* is a more recent industry and is practiced in France, Spain, Portugal, and Ireland in extensive systems, often combined with other shellfish culture. Crayfish *(Pacifastacus leniusculus* and *Astacus astacus)* are also cultivated in small amounts, and prawns *(Penaeus japonicus* and *P. kerathurus)* are being cultured in extensive or semi-intensive systems at an experimental level in France and Spain.

For the near future, Greece, Italy, Portugal, and France are expected to display the fastest growth in aquaculture production of all countries in the EC, an expansion that is mostly associated with the growth of sea bass, sea bream, and turbot production (which is expected to increase in France). Shellfish production is also expected to increase. Future production of salmon and trout will be determined primarily by marketing conditions. Catfish, carp, and mullet production currently has little market appeal. Overall, aquaculture production in the EC is expected to reach 966,000 metric tons in 1995, a 15 percent increase over the 1989 levels.

Canada

The aquaculture industry in Canada grew in part because of a strong commitment by federal and provincial governments, and a history of success in fisheries export and marketing. During the past 10 years, there has been a phenomenal growth in the commercial salmon farming industry, which is expected to continue for the next 10 years, although fluctuations in the market price for salmon make it a risky investment venture (Cook, 1990; Egan and Kenney, 1990).

Salmon, oysters, mussels, and marine trout currently dominate the aquaculture industry, but new species are in the research and development stage. Species expected to make a significant contribution to commercial aquaculture in Canada in the next 10 years include arctic char, bay scallops, nori, and Irish moss (Aiken, 1990). Research to develop new products is carried out by federal and provincial scientists, often in cooperation with private companies. A good example of this teamwork is the development of the commercial culture of Irish moss (*Chondrus crispus*) in Nova Scotia, based on collaboration between the National Research Council of Canada and Acadia Seaplants Ltd. (Isaacs, 1990).

During the 1990 World Aquaculture Society meeting in Halifax, Nova Scotia, the Minister of Fisheries and Oceans announced a government commitment to development of a world-class aquaculture industry in Canada in the 1990s. Implementation of this long-term strategy will be achieved by the support of science and technology, provision of an inspection system, assistance with market and commercial analysis, as well as advocacy and dialogue to promote sustained growth and development.

Shellfish

Two species of oysters are cultivated commercially: the Pacific or Japanese oyster *(Crassotrea gigas)* and the American or Virginia oyster *Crassotrea virginica*. A third species, the European (or Belon) oyster *Ostrea edulis* is under development in Nova Scotia. Oysters are produced by three methods in British Columbia: intertidal bottom culture, near-bottom culture, and off-bottom culture. Three years may be required to grow a marketable oyster on the bottom, but this can be cut to two years off-bottom. Suspension culture will produce more than 25 times the yield per unit area than can be obtained with bottom culture (Aiken, 1990). An estimated 3,900 metric tons of oysters were produced in British Columbia in 1989 (Price Waterhouse Management Consultants, 1990).

Mussel *(Mytilus edulis)* culture in Atlantic Canada has expanded considerably in the past 10 years, and in 1989 the five Atlantic provinces produced 3,137 tons of mussels valued at $5,520,000 (Muise, 1990). Cultivation developed in eastern Canada utilizes suspension technology (a long line system) to produce a premium quality product. Other shellfish considered for commercial culture in Canada are Manila clams *(Venerupis japonica)*; several species of scallops, including *Argopecten irradians* and *Patinopecten yessoensis;* and the pinto abalone *(Haliotis kamchatkana)* (Aiken, 1990).

Finfish

Finfish aquaculture registered a rapid growth in British Columbia: in 1976–1977 there were 5 farms licensed for marine trout and salmon, 51 freshwater trout sites, 2 carp sites, and 1 site licensed for carp and trout. In 1988, there were a total of 212 marine finfish sites and 149 freshwater trout sites (Price Waterhouse Management Consultants, 1990).

Rainbow trout has been reared since the 1950s in freshwater systems such as tanks, ponds, and raceways. Production increased by 10 percent yearly from 1976 to 1986 (when 100.8 metric tons of freshwater trout were sold commercially) (Price Waterhouse Management Consultants, 1990). More recently, rainbow trout has been raised in marine net-pen systems and sold commercially as "salmon trout."

Salmon

The British Columbia salmon farming industry has grown from 4 commercial farms in 1981 to 135 operating farm sites in 1989 and an estimated 120 sites in 1990 (Price Waterhouse Management Consultants, 1990). In 1989 the industry produced 12,385 metric tons of salmon with a landed value of Can $82.1 million (Egan and Kenney, 1990). Until 1986, coho *(Oncorhynchus kisutch)* was the dominant species cultivated (comprising 76 percent of the total farmed production that year), and the industry was dependent on government supplies of eyed eggs.

Since then, the industry has become self-sufficient in chinook salmon *(Oncorhynchus tshawytscha)* brood stock supply, and production of this species is now dominant (73 percent of the 1989 production), a shift attributed to early maturation and size problems related to coho production (Egan and Kenney, 1990). Atlantic salmon was first cultivated commercially in net pens in 1986 and has since increased to 8 percent of the total 1989 production. Its share is expected to increase further in the future, due to good growth and the higher price commanded in relation to the Pacific species (Egan and Kenney, 1990).

The salmon farming industry in eastern Canada is concentrated in the Bay of Fundy in New Brunswick, where approximately 49 salmon farms are currently operating. Together these farms have a combined estimated capacity of 8,500 metric tons (Price Waterhouse Management Consultants, 1990).

Latin America

The major aquaculture product in Latin America is shrimp. Shrimp farmers in the Western Hemisphere accounted for 11 percent of world production, or 61,000 metric tons in 1989. Ecuador produced 65 percent; countries producing about 5 percent each are Mexico, Honduras, Peru, and Colombia; Guatemala, Panama, and Brazil produced slightly less. Much of the production in the Western Hemisphere comes from extensive farms, but the trend is toward developing semi-intensive farming. *Penaeus vannamei* accounts for 92 percent of the production of farm-raised shrimp, which relies on wild shrimp for the production of seed stock. Disease represents the biggest obstacle to the future of shrimp farming in the Western Hemisphere. In the spring of 1990, Ecuador's $300 million a year industry was near collapse as a weather-induced disease epidemic struck its ponds and hatcheries (Rosenberry, 1990). It is not clear what the ultimate result of this epidemic will be, but production in mid–1990 had already been reduced by 40 percent; many of the ponds are not in operation, and no restocking is planned because of disease and reductions in seed stock availability.

Despite proximity to the U.S. market, the development of aquaculture in Latin American countries is slowed by government intervention, corruption, regulations, and permitting delays (often of one or two years)—"all the products of monumental bureaucracies" (Rosenberry, 1990). Recent changes in the political and economic environments in many of these countries (e.g., Mexico, Venezuela, Brazil, and Peru) are viewed as encouraging to the prospects for development of the shrimp farming industry.

INTERNATIONAL TECHNOLOGICAL DEVELOPMENTS

Issues of Concern

Issues faced by other countries include:

- pollution of coastal waters (particularly in Asia),
- shortage of coastal areas for expansion in Japan and many parts of the world,
- shrimp disease in Ecuador and Taiwan, and
- lack of governmental policies or institutional support for mariculture development.

Environmental impacts of aquaculture include:

- self-pollution through toxic and organic waste discharges;
- buildup of suspended solids;
- reduction in oxygen levels and introduction or augmentation of disease;
- habitat impairment and loss of natural resources;
- risks associated with transfers and introduction of exotics (of particular concern in Europe and the United States); and
- competitive use of resources, including land, water, and plant and animal resources.

Biological issues relate to biological and chemical unknowns in the areas of captive breeding, larval rearing, feed production for each life stage and appropriate to specific culture systems, and engineering problems in recirculating, closed or semiclosed, high-intensity culture systems.

Socioeconomic issues involve regulatory and administrative constraints, inadequacy of information and advice to investors, product marketing, and market saturation, to name but a few.

Culture systems and practices have been used to overcome constraints to marine aquaculture in other countries. Following is a summary discussion of these issues and their applicability to similar problems in the United States:

- Waste is being reduced by the development of methods to utilize properly the diets offered, to remove suspended solids efficiently, to incorporate water treatment and water reuse systems, and to collect waste for use as fertilizer.
- User conflicts over coastal resources are reduced by moving fish farms out into deeper water or to land-based facilities, or by zoning and clustering farms in selected sites.
- Member countries of the International Council on Exploration of the Seas (ICES) have adopted a Code of Practice to Reduce the Risks of Adverse Effects Arising From Introduction of Nonindigenous Marine Species.
- Attempts to circumvent disease and contamination of aquaculture products are carried out by individual countries. They include requiring health certificates for imports, as in North Sea countries for imported mussels; monitoring for toxic dinoflagellates and toxicity testing in Japan; requiring depuration of bivalves at specified centers in Spain; and maintaining strict quality standards with regular inspection in the Netherlands. Regulations generally pertain only to mollusks; they have been instigated in response to actual or potential health risks associated with eating these products.
- Research on controlled spawning to reduce aquaculture's dependence on wild seed, along with the development of larval rearing technology, is given priority in Japan, Norway, and many other countries.
- There has been some development toward improved feed quality to increase food conversion and to decrease waste (i.e., phosphorus).
- Some examples of solutions to socioeconomic constraints are the development of planned marketing strategies to promote future growth, government regulations that are conducive to growth in Canada (Price Waterhouse, 1990), and self-imposed restraints on 1989 salmon production in Norway in the face of a world market excess (Folsom and McFetters, 1990). In the latter case, the farming industry, through the Norwegian Fish Farm-

ers' Sales Organization (NFFSO), is taking strong action to shore up prices. The NFFSO plans to borrow $200 million to finance the purchase and freezing of 20,000–40,000 tons of salmon to keep it off the fresh fish market.

EXAMPLES OF AQUACULTURE POLICY IN OTHER NATIONS

Several nations have been successful in developing strong aquaculture enterprises that make significant contributions to the national economy. Although such experiences are not directly exportable to the United States because of different social, cultural, demographic, and economic conditions, they may suggest some fruitful avenues for U.S. action. Examples of nations that are most similar to the United States in culture and political organization are discussed below in order to maximize the potential applicability of any lessons that can be learned.

Canada

Status

Canada has a long aquaculture history dating back to turn-of-the-century oyster farms. Both the federal government and university structures have supported and encouraged expansion and development.

The British Columbia oyster industry began around the turn of the century and has been growing at a steady pace ever since. Clam aquaculture (primarily the Manila clam) is in the development stage, fueled by strong markets. Present efforts are targeting pseudofarming, which involves collecting wild spat and raising them on tidal flats under more optimal conditions. Mussel and scallop aquaculture research is also being spurred by strong markets but is hampered by culture problems.

Regulations

The Ministry of Agriculture and Fisheries administers the aquaculture lease program, which includes permits, licenses, and reviews. Aquaculture product licenses and permits include the aquaculture license from the Ministry of Agriculture and Fisheries, a municipal business license, a municipal sewage disposal permit, and a waste management permit. In addition, a shellfish transport permit is required. Site requirements include a federal water lot lease if operating on federal land and an occupation lease from the Ministry of Crown Lands. In addition, there may be local zoning ordinances.

Norway

Status

Of the many nations engaged in cold water fish farming, Norway is recognized as a leader in aquaculture development and production. Norwegian production of farmed salmon has risen from 4,000 tons in 1979 to 150,000 tons in 1990 (the 1990 figure represents a 25 percent increase compared to 1989). More than 90 percent of this production is exported. In fact, aquaculture is Norway's fastest growing industry, with an average annual growth rate of 47 percent from 1980 to 1986. This remarkable growth is attributable to good water quality; low but ideal sea temperature owing to the Gulf Stream; sheltered, ice-free sites behind a myriad of coastal islands; innovative technology; and the development of new markets. Currently, about 750 fish farms in Norway provide direct employment to 6,000 people, with another 9,000 jobs provided indirectly through educational services, research, and public administration. Although salmon aquaculture dominates Norwegian fish farming production, other cultivated species include trout, arctic char, oysters, and mussels. Currently, research is also directed at attempts to farm halibut, Atlantic cod, and ocean wolffish (Tilseth, 1990).

Legislation

Fish farming in Norway is regulated under the Fish Farming Act of 1985. The objective of the act is to ensure a balanced development of the industry and to make it profitable and viable. The act applies to both freshwater and saltwater aquaculture, and includes the handling and feeding of fish and shellfish, as well as the geographic allocation of new farms. Under the act, anyone wishing to enter the industry must first receive a license from the government, and since 1977 the government has limited the number of licenses issued. Norway restricts the number of fish farmers because of a desire to adapt production to market demand through balanced development. The limitation also is associated with the capacity of the nation's veterinary and extension services. The demand for licenses is strong, as was demonstrated in 1986 when 2,500 applicants competed for 150 new licenses. The number of hatchery and smolt operations, however, is not limited, and licenses for cultivating shellfish or other species of fish are granted more liberally. Norway also must consider the fact that aquaculture is growing rapidly in such places as British Columbia and Chile.

Internally, the outlook, although promising, will be hampered by the regulations described earlier that prevent Norwegian fish farmers from exploiting economy of scale in production and in the benefits of horizontal and vertical integration. An issue of increasing concern has to do with salmon escapes from fish farms. In 1989, 20 percent of the fish caught by fishers had escaped from fish farms. This situation has led to increasing concern about fish from farms breeding with and contaminating wild stocks. The Salmon Act of 1985 makes it illegal to move wild stocks from river to river.

Implications for the United States

The Norwegian experience in aquaculture illustrates what can be expected with the combined support of formal statutory guidance, intense research, and financial assistance. Of course, the strict regulatory model adopted by Norway may not be appropriate for the United States, but the Norwegian investment in research is consistent with the U.S. approach to agricultural research. In comparing the fish farming experiences of Norway and Canada, government financial assistance seems to be the common factor in assessing the success of aquaculture in these countries. The Norwegian experience also indicates that aquaculture growth needs to be balanced with market demand and that, like other commodities (at least in the short run), aquaculture growth has its limits.

The United Kingdom

Status

Aquaculture grew rapidly in the United Kingdom during the 1980s, with total production increasing from 7,000 tons in 1980 to 45,000 tons in 1989. Most of this output has come from salmon and trout production, which amounted to 18,000 and 16,000 tons, respectively, in 1988. (Scotland is the second largest salmon producer in the world after Norway.) The combined wholesale value of the 1988 aquaculture output was 100 million pounds, compared with the total value of all fish and shellfish landings by U.K. vessels of 400 million pounds. The industry provides employment for about 5,000 people and at least a similar number in downstream industries. Fish farming has been especially important in the highlands and western islands of Scotland because the industry provides employment in many isolated and economically depressed areas. Currently, 244 salmon farming businesses are registered in Scotland operating at 459 sites. There are also about 400 sites for raising trout. Other species farmed in the United Kingdom include oysters, clams, and scallops.

Legislation

Responsibility for aquaculture development in the United Kingdom rests primarily with the Ministry of Agriculture, Fisheries and Food, and the other territorial fisheries departments for Wales, Scotland, and Northern Ireland. Aquaculture legislative controls are directed toward the establishment and operation of fish and shellfish farms, including disease and movement controls, planning, water abstraction and discharge, and navigation.

The primary responsibility of the fisheries departments is stipulated under the Diseases of Fish Acts (1937 and 1983) and the Sea Fisheries (Shellfish) Act (1967). These laws are directed at preventing the introduction and spread of pests and diseased fish. All fish and shellfish operations in the United Kingdom are required to register with the appropriate fisheries department and to maintain records of fish movements. The Sea Fisheries (Shellfish) Act of 1967 also grants the exclusive right to cultivate oysters, mussels, cockles, clams, scallops, and queen conch in designated waters. Activities associated with the release of nonnative fish and shellfish into the wild, the use of pesticides, and the licensing of medicines are regulated under the Wildlife and Countryside Act of 1981, the Control of Pesticides Regulations of 1968, and the Medicines Act of 1968, respectively.

Planning

Aquaculture planning control rests with both local planning authorities and central government. Fish farms have to comply with the provisions of the Town and Country Planning Act, 1971, and the Town and Country Planning Act (Scotland), 1972. Also in accordance with EC Directive 85/337/EEC, environmental assessments must be undertaken for salmon rearing developments that are judged likely to have significant environmental effects. Effluent discharges from fish farms are controlled under the Water Act of 1989. Fish farmers are required to obtain consent to discharge their wastewater and to observe the standards set by the appropriate national river authority (river purification authorities in Scotland). The act also extends the need to obtain a water abstraction license for certain farms in England and Wales. In Scotland, water abstraction for fish farming is based principally on a common law right of riparian owners to use water in rivers and streams.

Marine-based fish farms are almost without formal planning control procedures, but their operations normally require the consent of, and a lease from, the Crown Estate Commissioners (CEC). The role of the CEC in planning and approving marine-based aquaculture projects is under review by the Agriculture Committee of the House of Commons.

Marine fish farms must obtain navigation consents from the Department of Transport to ensure that cages and other anchored equipment do not interfere with navigation of vessels. Under the Shetland County Council

Act and the Orkney Council Act, both of 1974, the Shetland and Orkney Islands councils have jurisdiction over the waters surrounding the islands and are responsible for issuing work licenses, which are necessary for establishing fish farms.

Financial Assistance

Financial assistance for fish farming in the United Kingdom is available from several sources. The first is the regional selective assistance program operated by the Industry Department under the Industrial Development Act of 1982. A second source is government agencies such as the Highlands and Islands Development Board, the Scottish Development Agency, the Welsh Development Agency, and the Council for Small Industries in Rural Areas. The EC also provides aid for the establishment, extension, and modernization of fish farms. Since 1978 the EC has assisted 125 projects in the United Kingdom at a total cost of 10 million pounds (mostly for salmon farms). To qualify for EC aid a project must also be in receipt of national assistance of at least 10 percent of the eligible costs. The EC has also given assistance to capital investments concerned with the processing and marketing of farmed fish, including processing, packaging, freezing, chilling, and storage facilities. In addition, national assistance has been made available directly for promotional and marketing initiatives by contributions to such bodies as the Scottish Salmon Farmers' Marketing Board and the British Trout Association and, indirectly, through Food From Britain and the Sea Fish Industry Authority.

Outlook

The outlook for aquaculture in the United Kingdom is dependent on the availability of suitable sites and growing conditions, the costs of fish meal, competition from other countries, disease, environmental considerations, and judgments as to the benefits of aquaculture compared to tourism and other coastal uses. Future growth also varies by species. Farmed salmon production could increase by more than 50,000 tons by the mid-1990s, with most of this expansion coming from existing farms or those already planned. With increasing pressure on coastal siting, offshore salmon farming is likely to develop. Trout production is also expected to increase, perhaps reaching 25,000 tons by the mid-1990s, but further growth is likely to be limited by the availability of adequate freshwater supplies. Mollusk production is expected to increase significantly. There are about 10,000 hectares of productive ground in estuaries and inlets. These lands could produce 15 tons of oysters, 30–50 tons of mussels, or 10–25 tons of clams a year.

Implications for the United States

Aquaculture development in the United Kingdom offers interesting insights, including both similarities and differences with the U.S. aquaculture experience. Despite the lack of formal statutory guidance in the United Kingdom for the development of aquaculture, this sector has developed into a significant industry. Domestic factors in the United Kingdom responsible for this growth include financial support, favorable growing conditions (i.e., water quality, temperature), and minimum resistance from other coastal users. The inherent international factor (i.e., international capital and economic competition) of the European environment has also had an important role in furthering the U.K. aquaculture industry. Continued expansion of the industry, however, faces challenges in siting, planning, and public acceptance. Yet, despite these difficulties, aquaculture in the United Kingdom has gained a more recognized status as a legitimate coastal area enterprise than it currently enjoys in the United States.

REFERENCES

Aiken, D. 1990. Commercial aquaculture in Canada. World Aquaculture 21(2):66-75.

Bettencourt, S.U., and J.L. Anderson. 1990. Pen-Reared Salmonid Industry in the Northeastern United States. Department of Marine Resource Economics, University of Rhode Island, Kingston, R.I. 147 pp.

Cook, R.H. 1990. Salmon farming in the Bay of Fundy—The challenge of the future. World Aquaculture 21(2):46.

Egan, D., and A. Kenney. 1990. Salmon farming in British Columbia. World Aquaculture 21(2):6-11.

Folsom, W.B., and B.D. McFetters. 1990 World salmon aquaculture. Proceedings of a Marine Technology Society Conference "Science and Technology for a New Oceans Decade" 623-628.

Food and Agriculture Organization (FAO). 1989. Fisheries Circular No. 815 Rev. 1.

Food and Agriculture Organization (FAO). 1990. A new definition of aquaculture. Fisheries 15(4):54.

Gousset, G. 1990. European eel (*Anguilla* L.) farming technologies in Europe and in Japan: Application of a comparative analysis. Aquaculture 87:209-235.

Isaacs, F. 1990. Irish moss aquaculture moves from lab to marketplace. World Aquaculture 21(2):95-97.

Juario, J.V., and L.V. Benitez, eds. 1988. Perspectives in Aquaculture Development in Southeast Asia and Japan. Aquaculture Department SEAFDEC, Tigbauan, Iloilo, Philippines.

Mito, I., and J. Fukuhura. 1988. Aquaculture development in Japan. Pp. 39-72 in Perspectives in Aquaculture Development in Southeast Asia and Japan, J.V. Juario and L.V. Benitez, eds. Aquaculture Department SEAFDEC, Tigbauan, Iloilo, Philippines.

Muise, B. 1990. Mussel culture in Eastern Canada. World Aquaculture 21(2):12-23.

Needham, T. 1990. Canadian aquaculture—let's farm the oceans. World Aquaculture 21(2):76-80.

Price Waterhouse Management Consultants. 1990. Long term production outlook for the Canadian aquaculture industry (1990 edition) an overview. Report prepared for Department of Fisheries and Oceans, Ottawa, Ontario.

Rosenberry, R. 1990. Shrimp farming in the Western Hemisphere. Presented at Aquatech 90, Malaysia, June.

Rosenberry, R. 1991a. World shrimp farming. Aquaculture Magazine (September/October):60-64.

Rosenberry, R. 1991b. World shrimp farming. 1991. Aquaculture Digest, San Diego, Ca.

Tilseth, S. 1990. New marine fish species for cold-water farming. Aquaculture 85:235-245.

Tseng, C.K. 1981. Commercial Cultivation. Pp. 680-725 in Biology of Seaweeds, C. Lobban and M. Wyne, eds. Berkeley: University of California Press.

Appendix B

Freshwater Aquaculture in the United States

OVERVIEW OF PRODUCTION

Of the roughly 0.3 million metric tons (mmt) of aquatic life grown for food in the United States, three-quarters or more are freshwater organisms. Most of the freshwater production consists of catfish, crayfish, and rainbow trout, in that order of importance. Large numbers of freshwater organisms are grown for purposes other than their immediate use for food. These include ornamental fish, baitfish, trout and other species stocked for recreational fishing, and salmon smolt released at sea for ocean ranching operations. In these latter applications, the fish are not normally sold, utilized, or accounted for by weight, so that it is difficult to assess their importance in U.S. aquaculture relative to the importance of the food species.

Catfish

Catfish farming is, far and away, the major aquaculture success story in the United States. Beginning no earlier than the mid-1960s, when a few thousand kilograms of channel catfish *(Ictalurus punctatus)* were grown in Arkansas, the annual crop had increased to more than 1,000 metric tons raised in 18 different states by 1969. By 1980, the center for the industry had shifted to the lower Mississippi delta, and national production had increased to more than 20,000 metric tons. Since then, catfish farming has grown by some 30 percent per year to a crop of 155,000 metric tons sold to processors in 1989, 90 percent of which is grown in the state of Mississippi (Rhodes, 1987).

Availability of fingerlings early in the year remains a major constraint to the industry. Only recently have efforts been made in the area of demand spawning through manipulation of photoperiod, temperature, and diet. Some stock improvement through selective breeding also has been initiated, together with modern genetic manipulation, such as gynogenesis, to produce faster-growing all male populations. A chronic problem in catfish culture is off-flavor, which may result in the rejection of entire crops.

The industry currently tests fish before harvesting and processing to ensure that off-flavor fish are eliminated before marketing. With the increase in production, the price of catfish dropped in 1989 from $1.72 to a low of $1.39 per kilogram (kg) of whole fish at pond side, suggesting that the market might be saturated. However, creation of a marketing association in October 1989 resulted in a price rise back to $1.72 by March 1990 while production continued to increase. At that price, the current annual crop has a value to the grower of some $265 million (Huner, 1990a, 1990b).

Crayfish

The inhabitants of Louisiana, many of whom are of French background, have long harvested and enthusiastically consumed native populations of crayfish (locally "crawfish"), a food virtually ignored by consumers in much of the United States but highly popular in parts of Europe. Louisianians have also, for many years, practiced a simple form of aquaculture for the most popular and abundant species, the red swamp crayfish *(Procambarus clarkii)* and, to a lesser extent, the white river crayfish *(P. blandingi)*.

Crayfish culture is an extremely nonintensive, low-cost form of aquatic farming that is barely distinguishable from capture fishing. Shallow (<1.0 meter [m] wetland ponds (often rice paddies) are stocked (usually just once) with adult animals at 50 to 100 kg per hectare. They often are planted with starter populations of several species of edible plants (if rice is not present in abundance), such as alligator grass *(Alternanthera phylloxeroides)* and water primrose *(Jussiaea* spp.*)*. However, the crayfish eat almost any soft aquatic or terrestrial plant and often will thrive on natural vegetation.

After mating occurs in late spring, the ponds are slowly drained. The animals then dig and move into burrows in the soft mud bottoms, where they remain until egg laying occurs in early fall. At that time, the ponds are refilled, the young hatch from the eggs, and growth proceeds until the next mating season. Once ponds are initially stocked with crayfish and, if necessary, planted with vegetation, the aquaculturist's role is merely to manipulate the water level of the pond.

Harvesting is by trapping, which is rather primitive and inefficient, and represents the most costly aspect of the culture operation, but which may be done throughout the flooded period (i.e., from November to May, with highest production from March to May). Traps are traditionally baited with fish wastes. Artificial baits are now formulated by some of the large feed manufacturers, but their use is economically marginal. Crayfish and rice are often grown in rotation in the same pond, but care must be taken to guard against the use of pesticides, to which the crustacea are highly sensitive.

During the 1970s, about 1,000 metric tons of crayfish were reared on 7,000 hectares of ponds in Louisiana. Since then, the industry has expanded rapidly in response to increased local and international demand, the latter exacerbated by the decimation of European crayfish populations by disease. By the mid-1980s, farms had spread into neighboring Texas, Mississippi, and Florida, and yields approached 50,000 metric tons from 40,000 hectares of ponds.

In danger of overproduction, the industry increased marketing efforts, with the result that the more popular local "Cajun" and "Creole," as well as the more traditional French, recipes are now offered by specialty restaurants throughout the United States. However, continued expansion of the industry (53,000 hectares of ponds in Louisiana alone in 1988–1989) eventually led to market saturation and a drop in price to $0.77/kg by 1990, which has led to a decrease in intensification of the farming (i.e., trapping) effort and a corresponding decrease in yield per unit area, with an overall current yield of approximately 30,000 metric tons by the industry and a value to the grower of $24.5 million (Huner, 1990a, b).

A new product within the past few years is newly molted soft-shelled crayfish, comparable to soft-shelled crabs both in method of production and in utilization (Huner, 1990c). Because of present marketing problems with traditional crayfish crops, there was a rush by growers to the production of soft-shelled crayfish in 1988–1989, which unfortunately quickly saturated the new undeveloped market for that product. The future of the U.S. crayfish culture industry is therefore currently in a state of some uncertainty (Catfish News, 1989).

Trout/Salmon

Trout culture is undoubtedly the oldest form of freshwater aquaculture in the United States, having been introduced from Europe more than 100 years ago to provide or enhance sport fishing in both private and public waters. Today, more than 200 million trout of several species are reared in some 350 state and federal hatcheries for distribution to public waters for sport fishing.

An additional 1 billion (approximately) Pacific salmon are reared to the size of smolt (the physiological condition at which they may be introduced to salt water, a stage that varies in size and age with the species of salmon) in federal and nonprofit private hatcheries in the Pacific Northwest for subsequent release to the sea. This program is carried out as mitigation for the loss of natural spawning areas in rivers through dam construction or, more recently, simply as enhancement of natural reproduction. Return of stocked smolt as mature salmon generally averages 2–5 percent, a figure that has increased with the production of larger, healthier smolt stock and now accounts for a significant fraction (about 30 percent or more) of U.S. salmon landings.

Such hatchery operations, particularly at the federal level, have provided the principal basis for development of the technology of salmonid culture. This has now become routine, and no striking new advances in the field have been made recently, but there has been steady improvement of diet and reproductive control, as well as some significant stock improvement through selective breeding.

Trout and salmon normally inhabit cool, clean water, and their cultivation, therefore, depends on the rapid exchange of flowing temperate water (10–20°C) of high quality. Eggs and sperm usually are stripped artificially, and eggs are incubated and fry reared to fingerlings in flowing water systems within hatcheries. Fingerlings are then traditionally reared outdoors to the desired size in concrete raceways, although other grow-out systems have been used successfully. The fish accept artificial formulated feed as soon as the yolk sac is absorbed, and appropriate formulations have been developed for each stage of their growth and development.

Although several species of trout are hatchery reared for sport fishing, the cultivation of trout for food has been restricted largely to the rainbow trout *(Salmo gairdneri* now called *Ancorynchus mykiss),* native to the U.S. Pacific Coast. Farming of the species for food was first begun in the 1950s in the state of Idaho, primarily owing to the advantage of a bountiful supply of cold, clean running spring water issuing from the Rocky Mountains in the Snake River valley. Although trout are now grown for food in virtually every state with a suitable climate for their cultivation, 76 percent of the U.S. production of some 25,000 metric tons still comes from Idaho.

The industry has grown slowly over the past decade, increasing by approximately 5 percent per year, on average. The principal constraint to more rapid growth is the relatively low price of the product, currently about $2.40/kg, which is set by competition from Japan and Europe but is marginal for a species that requires high-quality feed and relatively intensive culture technology. Traditionally, the product has been a portion-size (340 grams, or 3/4 pound, 12 inches) fish, fresh or frozen, gutted, and head-on. A trout of that size used to be grown in Idaho in about 18 months but now takes just over a year from the egg. Increasingly, the fish are now marketed fresh, and more than 50 percent are boned and sold as fillets. Further, an increasing fraction is being grown to a larger size of 1 kg or more, is pigmented a vivid orange color, and is marketed fresh, presumably to compete with marine cage-grown salmon and trout.

Bait and Ornamental Fish

Goldfish *(Carassius auratus)* were introduced to the United States from Japan in the late nineteenth century and have been reared in this country for nearly 100 years in a "goldfish belt" extending from Maryland to Arkansas. Initially bred exclusively for the ornamental or aquarium trade, the smaller and less attractive specimens began to be sold for bait and as feed for carnivorous ornamental fish by the middle of this century. Today more "feeder" than ornamental goldfish are reared and sold, but the value of the latter is still greater (Martin, 1983). At about the same time that goldfish began to be used for bait, other more traditional bait, particularly the golden shiner *(Notemigonus crysoleucas)* and the fathead minnow *(Pimephales notatus),* which belong to the same "minnow" family as goldfish, began to be reared in the central part of the country, particularly Arkansas.

All of the above fish are reared in much the same way in shallow, earthen ponds ranging in size from 0.1 to 25 hectares through which spring water is circulated slowly. The fish spawn naturally along the pond banks in the spring, either on grass or on artificial spawning mats made of sphagnum moss or other fibrous material. In the latter case, the mats are usually transferred to fishless breeder ponds after the adhesive eggs are laid, to prevent cannibalism. Grow-out takes 6 to 18 months, depending on the size of fish desired (Martin, 1983).

The tropical ornamental fish industry is centered in Florida, which provides more than 90 percent of the aquarium fish, exclusive of goldfish, sold in the United States. The industry consists of two kinds of operations: (1) importation of exotic species from tropical Southeast Asia, Africa, or Latin America for resale after holding and growing-out for various periods of time; and (2) maturation, breeding, and rearing to marketable size of a number of exotic species. Only about one-third of the sales by Florida growers currently belong to the second category of fish grown throughout their life cycle on the farms, and these represent 10 percent or less of the total number of species handled. There are about 12 species groups in that category, of which 5 (the live-bearing species) make up 70 percent of all sales. However, an increasing number of the more difficult to rear egg-bearing exotics have been brought under reproductive control by some progressive farmers in the past decade.

The fish are grown, for the most part, in small earthen ponds averaging about 200 m^2 in area, usually below the water table, so that they must be pumped dry for cleaning and water exchange. In hot weather, they may require aeration and, in cold weather, continuous pumping of deep, relatively warm well water and/or covering of the surface. The fish are fed prepared rations that differ considerably in formulation and amount applied. Antibiotics, drugs, and chemicals are used as needed for disease, pest, and weed control, but practices tend to be empirical and are far from standardized in the industry.

A study by the Florida Game and Freshwater Fish Commission estimated that the state's tropical fish industry consisted of some 215 full-time fish farmers utilizing an area of 433 hectares with an annual value to the growers of $26 million in sales (Knox and Drda, 1982–1983). Initially, the bait and goldfish industries were similarly fragmented into small family units but, over the past 25 years, have gradually become consolidated into fewer larger farms. About 10 growers dominated the fancy goldfish industry in the 1960s, and one large company has since captured about 80 percent of the business. The larger feeder goldfish and bait minnow industry is represented by some 50 growers, of which one-half dozen dominate.

Documentation of the size and monetary value of these nonfood fish culture industries in the United States is not well documented. One estimate put "baitfish" production at 12,200 metric tons worth $56 million in 1987, but it is not clear what the category included. An experienced academic and commercial goldfish culturist estimated that industry to be worth $10 million to $20 million to the farmer (Martin, 1983). It would thus appear that the entire category of nonfood fish culture has an annual value to the grower of $50 million to $100 million, making it one of the economically more important forms of U.S. aquaculture.

Alligators

Alligators are valuable for their hides, worth up to $50 per linear foot on the international market, and for their meat ($15/kg on the specialty food market). A 4- to 5-foot animal can be grown from an egg in about 14 months and brings about $20 to $30 per foot (in late 1990); wild animals are more valuable and are priced at $40 to $50 per foot..

There are presently some 180 alligator farms, of which 90 are in Louisiana and 40 are in Florida. Eggs are captured from wild nesting animals under a carefully controlled licensing arrangement, which includes the proviso that 17 percent of the reared stock be returned to the wild when they reach 4 feet in length. The animals are grown in heated incubators and fed rations prepared on-site from nutria, large swamp-dwelling rodents. Commercially produced pelleted feeds are also available in several sizes. A 5-foot alligator costs $50 to $75 to raise. Currently, about 100,000 cultured alligators are marketed per year, making this fledgling industry worth some $15 million to $20 million (Fish Farming International, 1990).

Hybrid Striped Bass

Hybrids of the anadromous striped bass *(Morone saxatilis)* female and the freshwater white bass *(M. chrysops)* male have been hatchery reared since the 1960s for sport fishing and forage fish control in large southern U.S. lakes and reservoirs. For the past decade, research has been carried out on the feasibility of growing the hybrid as an aquaculture food crop. These research and development efforts have included semi-intensive pond culture in North and South Carolina and elsewhere along the eastern seaboard, and highly intensive tank culture in the California desert using geothermal well water. The hybrid can be grown in either salt or highly alkaline fresh water, the latter having most frequently been used in the initial efforts. Several of these efforts have now achieved commercial status, including a new intensive tank culture project in Mississippi. However, the industry is still quite small. In 1987, approximately 450 tons were marketed, of which 75 percent were produced by one of the intensive tank farms in California.

The industry is still dependent on the female striped bass hybrid and the male freshwater white bass hybrid, but it has not been able routinely to produce sexually mature hybrids that spawn in captivity, relying instead on hormonally induced spawning of ripe-running wild females. The latter are difficult to acquire, both legally and logistically, and the supply of young is restricted to one brief period of the year.

The intensive culture method, involving heavy application of high-protein feed, pumping large volumes of water, the use of liquid oxygen, and a high capital investment for tanks and equipment, imposes a break-even cost of $4 to $5/kg for the product to the grower. If wild stocks of striped bass return to the East Coast in abundance, as appears now to be the case, the cost of rearing the hybrid striped bass in captivity must be lowered significantly, perhaps by the use of less intensive pond culture technology, for the industry to prove viable (Van Olst and Carlberg, 1990).

Tilapia

Several species of the genus *Tilapia (oreochromis)*, native to the lakes of Africa, have now spread to tropical and semitropical habitats throughout the world, including the southern United States. Herbivorous or omnivorous in its feeding habits, *Tilapia* can be grown inexpensively in extremely dense culture and with high yields. Not surprisingly, it has become a major target for aquaculture throughout the tropical world.

A major constraint to tilapia culture is its early sexual maturation and extreme fecundity, resulting in rapid overpopulation and stunting in ponds and impoundments where it is stocked. The problem has recently been partially resolved by hybridization of tilapia species to produce sterile or monosex (all-male) offspring. Selective breeding of the hybrids has also resulted in red varieties that are more attractive and marketable than the dark-colored native species. It is now possible to produce red hybrid tilapia of 0.5 to 1.5 kg in extremely dense, highly intensive tank culture. However, to find market acceptance of the species in this country has proved difficult at a price that is economically viable, and unlike its success elsewhere in the world, tilapia farming in the United States has yet to become an established industry.

Macrobrachium

The large freshwater shrimp or prawn, *Macrobrachium rosenberghii,* is grown in small commercial operations in Puerto Rico and Hawaii. With the exception of certain behavioral characteristics, such as dominance and territorialism in large males (which can be at least partially alleviated by frequent selective harvesting), there are no particular problems or constraints to its cultivation. However, the young must be reared in salt water prior to transfer to fresh water for adult grow-out, a factor that limits the places where it can be reared. In addition, the animal is truly tropical in its habitat requirement, so that year-round growth in most of the United States is limited. A number of *Macrobrachium* farms were started up in Hawaii over the past decade, but most have since gone out of business or have converted to marine (penaeid) shrimp. Marketing problems have hindered commercial success with the species.

Sturgeon

The white sturgeon *(Acipenser transmontanus)* is another anadromous fish that grows equally well in salt water or hard fresh water. It has been cultured primarily in California, where the industry has expanded rapidly over the past five years, and the product has been marketed routinely at attractive prices. To date, the culture system of choice is intensive tank systems using large cylindrical fiberglass tanks supplied with flow-through or recirculated fresh water. Pond culture systems have not been particularly satisfactory for sturgeon.

Major problems encountered by the industry include disease and the reliance on wild female brood stock. Diseases have been a recurring problem, and identification of control measures is a priority research area, along with development of domesticated brood stock. Mature males can now be produced routinely in captivity, and some cultured females (approximately six to seven years old) have recently exhibited ovarian development. Maturation of females in captivity could lead to an aquaculture-based caviar industry.

Experimental culture of the Atlantic shortnose sturgeon *(Acipenser brevirostrum)*, an endangered species, has developed to the point that fingerlings are produced regularly and are stocked in natural waters. Overall, sturgeon culture is expected to continue to expand, but with most grow-out operations located in freshwater, rather than saltwater, environments.

REFERENCES

Catfish News. 1989. Louisiana Florida slug it out on crawfish. (May/June):3.

Fish Farming International 1990. Alligator earnings boost U.S. farms. 17(5): 21.

Huner, J.V. 1990a. Crawfish and catfish in Louisiana. Fish Farm. Int. 17(1):28, 30.

Huner, J.V. 1990b. Aquaculture worth over $200 million to Louisiana. Fish Farm. Int. 17(4):20.

Huner, J.V. 1990c. New horizons for the crawfish industry. Aquaculture Magazine 16(5):65-70.

Knox, R.H., and T.F. Drda. 1982–1983. Florida Aquaculture Survey, 1982–1983. Florida Game and Freshwater Fish Commission on Aquaculture Project. 54 pp.

Martin, M. 1983. Goldfish farming. Aquaculture Magazine 9(3): 38-40; 9(4): 38-40; 9(5): 30-34.

Rhodes, R.J. 1987. Status of world aquaculture. Aquaculture Magazine 17th Ann. Buyers Guide, p. 8.

Van Olst, J.C., and J.M. Carlberg. 1990. Commercial culture of hybrid striped bass. Aquaculture Magazine 16(1):49-59.

Appendix C

Federal Marine Aquaculture Policy

Numerous federal and state agencies regulate and promote aquaculture. However, four agencies of the federal government—the U.S. Department of Agriculture (USDA), the National Marine Fisheries Service (NMFS), the Sea Grant College Program in the Department of Commerce, the U.S. Fish and Wildlife Service (FWS) in the Department of the Interior, and the National Science Foundation (NSF)—play major roles in various aspects of marine aquaculture development. Estimating total federal expenditures by these four agencies for marine aquaculture is problematic because marine aquaculture expenditures can be difficult to identify. For example, aquaculture expenditures do not always differentiate between freshwater and marine programs, as in the case of funding for the USDA National Aquaculture Information Center. Another example is that the FWS has received no appropriations under the National Aquaculture Act of 1980, or under subsequent amendments, but operates hatcheries and conducts marine aquaculture research under other authority. For these reasons, the figures quoted below should be viewed as a best estimate of total federal marine aquaculture expenditures. Current marine aquaculture activities of these four agencies are detailed below.

UNITED STATES DEPARTMENT OF AGRICULTURE

The USDA carries out much of its aquaculture-related programs under the Assistant Secretary for Science and Education through the Cooperative State Research Service, the Agricultural Research Service, the Extension Service, and the National Agricultural Library.

The Cooperative State Research Service (CSRS) works with state agricultural experiment stations, forestry schools, the land-grant colleges, Tuskeegee Institute, and colleges of veterinary medicine. CSRS allocates formula funds to states for maintaining high-quality research programs in aquaculture. CSRS awards grants in aquaculture on a competitive basis through the Aquaculture Special Grant Program and the National Research Initiative (formerly the Competitive Grants Office). CSRS also administers special grants in aquaculture on research problems that Congress believes are important to the nation. In cooperation with the Extension Service, CSRS provides funding for research and extension activities through the five Regional Aquaculture Centers. Additionally, research in the private sector is supported through the Small Business Innovation Research (SBIR) Program.

The Agricultural Research Service's aquaculture research program includes marine shrimp in the Pacific region; cold freshwater species in the Northeast; and warm freshwater species in the mid-South region of the United States. Research is conducted on genetics, breeding, nutrition, disease diagnostics and control, water quality and use, and production systems to increase production capacity and technology transfer. Improved product quality and marketing are supported with research on processing, off-flavors, food texture and taste, packaging, food safety, and value-added products. Research programs are conducted at Stoneville, Mississippi; Southern Regional Research Center, New Orleans, Louisiana; Animal Parasite Research Laboratory, Auburn, Alabama; University of Hawaii, Honolulu, Hawaii; and Lane, Oklahoma. Cooperative research programs are conducted at the Oceanic Institute, Waimanalo, Hawaii, and the Spring and Groundwater Resources Institute, Shepherdstown, West Virginia.

The Extension Service is the primary educational arm of USDA. Through the Cooperative Extension System's 74 land grant universities located in all 50 states, 6 territories, and the District of Columbia, programs are implemented in parternership with federal, state, and county levels of government. The system functions as a nationwide educational network and includes professional staff in nearly all of the nation's counties. It provides for the transfer of research, technology, and management information through educational programs and technical assistance. The Cooperative Extension System relies on information generated by research and helps interpret research results to speed the application and dissemination of this information to the public.

Many states have developed extension educational and service programs in aquaculture. They provide workshops for new fish farmers, short courses in management and fish diseases, aquaculture demonstrations, farm visits, field days, in-service training programs, and 4-H youth programs; they also distribute written and videotaped educational materials to assist aquaculture development. Many educational programs include newsletters on a variety of aquaculture topics.

The National Agricultural Library (NAL) has a National Aquaculture Information Center (NAIC), which was mandated by the National Aquaculture Improvement Act of 1985 to serve as a repository for national aquaculture information. The library acquires materials through purchases, gifts, and exchanges, including books and journals, microfiche collections, audiovisual material, and computer software in the field of aquaculture. It also provides document delivery service for many of these materials through its Lending Branch.

Staff of the NAIC publish bibliographies of interest to potential and practicing aquaculturists; conduct on-line and CD-ROM (compact disc, read-only memory) computerized searches of aquaculture-related data bases, and provide general information, bibliographies, and referrals to aquaculture extension specialists or other contact sources. The center networks with states, Regional Aquaculture Centers, libraries, and the public and private sectors to enhance information exchange in the nation. Staff of the Aquaculture Information Center at NAL are members of the federal Joint Subcommittee on Aquaculture and its supporting Information Task Force. They place emphasis on clearinghouse activities, and explore and utilize new computer technologies to improve information collection and exchange in the field.

Table C-1 provides a breakdown of USDA expenditures on aquaculture. A reasonable estimate for marine aquaculture expenditures is approximately one-half of the total CSRS amount (i.e., $6,643 million in 1990) (Meryl Broussard, CSRS/USDA, personal communication, 1991).

USDA Regional Aquaculture Centers

The USDA carries out much of its aquaculture-related programs through five Regional Aquaculture Centers. The Regional Aquaculture Centers are administrative entities that focus on the needs of aquaculture in the United States on a regional basis. The centers identify research priorities and fund regional and cooperative research and educational extension programs in aquaculture. Center programs strive to complement competitive grants programs already provided by the USDA, the Sea Grant College Program, and other public institutions. A variety of public and private nonprofit institutions participate in center programs.

The objectives of the regional centers are to:

- act as a mechanism for assessing needs, establishing priorities, and implementing regional research and extension programs in aquaculture;

TABLE C-1 USDA Aquaculture Programs (thousand U.S. dollars)

	1988	1989	1990
Agricultural Research Service			
Feeds (Hawaii)	905	892	1,332
Breeding and Genetics (Stoneville, Miss.)	486	506	497
Off-flavor (New Orleans, La.)	644	804	789
Cage Culture (Oklahoma)	264	275	494
Water Quality (West Virginia)		654	1,090
Diseases and Parasites (Auburn, Ala.)	61	65	63
Total	2,360	3,196	4,265
Cooperative State Research Service			
Hatch Act	1,414	1,669	1,664
McIntyre-Stennis	38	37	37
Animal Health, section 1433	115	147	145
Evans-Allen	1,155	1,317	1,354
Special Grants	1,480	1,705	2,304
Aquaculture Centers	3,500	3,750	3,703
Competitive Grants	—	424	424
Direct Federal Administration	2,236	2,736	3,654
Total	9,938	11,785	13,285
Extension Service			
Smith-Lever		1,491	1,491
Total		1,491	1,491
National Agricultural Library			
Aquacultural Information Center		247	307
Total		247	307
Total, Science and Education	12,298	16,719	19,348

SOURCE: Personal communication. Meryl Broussard, Cooperative State Research Service, USDA. (1991).

- provide a nucleus for information exchange between the aquaculture industry and researchers;
- facilitate administration and implementation of cooperative regional research and extension programs in aquaculture;
- provide research and extension linkages in the region;
- complement and strengthen existing research and educational extension programs in public and private institutions; and
- help coordinate interregional and national programs.

Each center has a board of directors, an industry advisory council, and a technical committee to assist in identifying industry's needs and to ensure that the research undertaken is relevant and has technical merit. Members of the board of directors of a center are selected from within the region and from participating institutions. The board is responsible for overall administration and management of the regional center program, allocating fiscal resources and developing priorities for regional aquaculture and educational extension activities and projects. The industry advisory councils are composed of representatives of state and regional aquaculture associations; they include people from supply industries, aquaculture production, marketing, processing firms, and financial institutions. They identify problems and needs of the industry in the region, and recommend research and extension priorities from an industry perspective. The technical committees develop recommendations from research and extension perspectives. The industry advisory councils and technical committees work closely together and jointly review proposed projects and the progress of current projects. All proposals undergo external review for technical merit and industry application.

The research and extension projects supported by the centers address a diversity of issues constraining the industry's profitability, including reproduction, genetics, nutrition, alternative sources of protein for fish feeds, disease, predation, waste management, regulations affecting the industry, economics and marketing, development of new technology, technology transfer, and information exchange. Research projects include species of current economic importance (including oysters, salmonids, catfish, and crayfish) and species showing promise for future production (including hybrid striped bass, sturgeon, tilapia, walleye, shrimp, and giant clams). A review of the research expenditures over the first four years of the Regional Aquaculture Centers indicates that marine, anadromous, and freshwater species each have been the subject of about one-third of the research and extension effort. Table C-2 summarizes funding for aquaculture research by the five

TABLE C-2 Funding Provided by USDA Regional Aquaculture Centers, FY 1988–1990 (U.S. dollars)

Regional Aquaculture Center	1988	1989	1990
North Central	398,348	495,993	330,866
Northeastern	625,401	591,775	429,874
Southern	826,910	758,426	404,664
Tropical and Subtropical	501,945	542,684	578,098
Western	578,500	724,406	360,526
Total	2,931,104	3,113,284	2,104,028

SOURCE: Regional Aquaculture Centers (1990).

Regional Aquaculture Centers during the period fiscal year (FY) 1988–1990.

The centers have been in operation less than four years so their effectiveness has yet to be assessed. Projects appear to be directed toward the most pressing industry problems and toward opportunities that can enhance profitability.

NATIONAL OCEANIC AND ATMOSPHERIC ADMINISTRATION

The National Oceanic and Atmospheric Administration (NOAA) involvement in marine aquaculture is provided through two programs—the National Marine Fisheries Service and the National Sea Grant College Program.

National Marine Fisheries Service

The National Marine Fisheries Service (NMFS) traces its involvement in aquaculture to the creation of the U.S. Commission of Fish and Fisheries in 1871. Initially, most of the commission's efforts were directed toward anadromous species, restoring the shad and Atlantic salmon and introducing chinook salmon to new environments. However, before the turn of the century, some culturing methods were developed for marine species such as cod, haddock, and herring. Eggs of these species were fertilized and hatched, and the fry were released. The aquaculture activities of the commission continued along these same lines until it was merged with the Biological Survey in 1940 to create the Fish and Wildlife Service (FWS).

At the time of the creation of FWS, the primary "marine" species cultured on a production basis were Columbia River salmon. This activity was the result of the passage of the Mitchell Act (1937), which required the construction of mitigation hatcheries to replace the spawning grounds that were lost by the building of dams on the Columbia River. From 1940 to 1971, the FWS also developed improved techniques for the culture of oysters, mussels, shrimp, catfish, and trout. In 1970, all of the marine elements were transferred out of FWS to create NMFS, including the responsibility of administering the 24 "Mitchell" hatcheries and rearing facilities. Presently all but one of the Mitchell facilities are in operation. Annual production is 100–120 million smolts. Approximately 75–80 percent of the production is of fall chinook salmon, followed by coho salmon, spring chinook salmon, steelhead trout, and cutthroat trout. Expenditures for these facilities in FY 1991 amounted to approximately $9.5 million (James Meehan, National Marine Fisheries Service, personal communication, 1991).

From 1971 until 1982, NMFS conducted a number of studies at various research facilities around the country aimed at assisting industry. Notable projects included oyster culture techniques, with particular emphasis on control of pathogens, at the Milford, Connecticut, and Oxford, Maryland, laboratories; salmon pen rearing techniques at the Manchester, Washington, field station; development of a disease-free Atlantic salmon brood stock, also at Manchester; and shrimp culture techniques at the Galveston laboratory.

NMFS aquaculture efforts were redirected in 1983 away from the goal of producing food to the management of common-property resources and endangered species. NMFS objectives for aquaculture became:

- to support and/or contribute to management objectives defined in fishery management plans developed under the Magnuson Fishery Conservation and Management Act or the interjurisdictional coastal fisheries program in cooperation with states;
- to contribute to the restoration and protection of endangered species or stocks under programs authorized by the Endangered Species Act; and
- to respond to Indian treaty obligations, legislative mandates, and court orders.

NMFS continues to disseminate aquaculture-related information and technological advances gained from its fisheries research. It also promotes the development and expansion of international markets for U.S. aquaculture industry products. In 1990, NMFS initiated a review of its aquaculture policy to determine its proper role in light of recent concerns about marine aquaculture in the United States. Table C-3 details NMFS FY 1991 aquaculture expenditures and the variety of programs that are supported.

National Sea Grant College Program

The National Sea Grant Program currently consists of 26 Sea Grant College Programs and three Sea Grant institutions in the coastal and Great Lakes states. The Sea Grant College Program combines research, education, technology transfer, and public service in a university-based program designed to enhance and promote the wise use of Great Lakes and marine resources.

The Sea Grant marine advisory program estimates that 12 percent of its effort is spent on aquaculture-related activities. No set figure is assigned to any research area in Sea Grant, and states are free to devote as much of their budgets to aquaculture-related topics as they decide is necessary to have a balanced program. In recent years, almost all of the programs have funded aquaculture-related projects, between 94 and 121 projects each year since 1979. All projects selected for funding are subject to outside peer

TABLE C-3 NMFS Aquaculture Expenditures, FY 1991 (dollars)

Saltonstall/Kennedy Grant Program	
Augmented roe production	77,600
Salmon farming health management program	121,500
Reduction of hatchery and aquaculture diseases	62,000
Sponge aquaculture feasibility study	19,400
Trochus reseeding	11,800
Regional yield trials for giant clam species	96,200
Commercial sponge aquaculture training	23,900
Rehabilitation and construction of oyster reefs	29,700
Beneficial environmental impacts of net-pen salmon culture	249,700
Domestication and mass culture of summer flounder	148,600
Economic production of microalgal aquaculture feeds	5,400
Subtotal	845,800
Hatchery biology—Columbia River salmon (includes 1,971,300 reimbursable)	2,922,600
Hatchery biology—Alaska chinook salmon	596,200
Kemp's ridley sea turtle recovery	342,500
Mahimahi aquaculture	188,000
International trade in U.S. aquaculture products	10,000
Aquaculture policy	20,000
Development of fish feeds	135,000
Seagrass culture	224,800
Aquaculture support—Newport, Ore. facility	327,900
Shellfish pathology	288,700
Subtotal	5,055,700
Accumulated total	5,901,500
Catfish aquaculture research (Stuttgart, Arkansas) (Pass-through to FWS)	2,750,000
Subtotal	2,750,000
Accumulated total	8,651,500
Mitchell hatchery O&M, fishways and stream screening	9,597,400
Subtotal	9,597,400
Accumulated total	18,248,900
Pacific salmon treaty (U.S.-Canada) and salmon enhancement	12,325,800
Columbia River dam damage passage problems	1,665,700
Ecological effects of dams (includes 1,981,300 reimbursable)	3,230,800
Subtotal	17,132,300
Total	35,381,200

SOURCE: Personal communication. James Meehan, NMFS, NOAA (1991).

TABLE C-4 Sea Grant Aquaculture Projects and Funding, FY 1979–1990 (thousand U.S. dollars)

Year	No. of Projects	Federal	Matching	Total
1979	98	3,707	3,293	7,000
1980	111	4,277	2,994	7,271
1981	115	3,735	2,744	6,479
1982	106	3,050	2,047	5,097
1983	109	3,914	2,778	6,692
1984	97	3,739	2,994	6,733
1985	121	4,414	3,721	8,069
1986	110	4,550	3,263	7,813
1987	121	4,544	3,111	7,655
1988	103	4,102	3,088	7,190
1989	94	3,640	2,861	6,501
1990	102	4,120	2,956	7,076

SOURCE: Personal communication. James McVey. Sea Grant College Program, NOAA (1991).

review. Matching funds are provided by the states, bringing the average yearly support for aquaculture to approximately $7 million as shown in Table C-4.

The National Sea Grant College Program has been a major contributor to the development of aquaculture technology for marine, estuarine, and Great Lakes species for more than 20 years. This activity was reinforced in the National Aquaculture Act of 1980 and the National Aquaculture Development Plan of 1983. The National Sea Grant Program has prepared its own aquaculture plan through a cooperative effort of involved institutions and has set priorities in line with the National Aquaculture Plan. Yearly guidelines are provided to Sea Grant Programs to focus research proposals on identified priority areas. The Sea Grant Program also supports workshops and symposia for key aquacultural species groups to help establish the industry and to focus on research needs to support the developing industry. Aquaculture projects funded by the Sea Grant Program from 1979 to 1990 are listed in Table C-4, along with federal and matching funds.

FISH AND WILDLIFE SERVICE

The mission of the FWS is to provide federal leadership to conserve, protect, and enhance wildlife and their habitats for the continuing benefit of people. This mission has been defined with respect to fisheries as a goal to promote and enhance conservation of the nation's freshwater, anadromous, and intercoastal fishery resources for maximum long-term public benefit.

Specific responsibilities of FWS include restoration of depleted nationally significant fishery resources, mitigation of fishery resources impaired by federal water-related development, assistance with management of fishery resources on federal and Indian lands, and maintenance of a federal leadership role in the scientifically based management of national fishery resources.

The FWS operates a system of fish hatcheries, fish health centers, fish technology centers, and fishery research centers. The aquaculture mission and activities of FWS are derived from its fisheries mission and the infrastructure that supports its fisheries activities. The mission of FWS in private aquaculture includes two major components: (1) provision of information, technical assistance, and services to the aquaculture industry; and (2) assisting the aquaculture industry to ensure that its development takes place in a manner consistent with responsible resource stewardship.

Freshwater finfish receive the greatest emphasis in the FWS aquaculture programs and activities; however, anadromous, estuarine, and coastal migratory species receive considerable attention as well. More than 20 of the 78 FWS national fish hatcheries rear species that spend at least part of their life in salt water. In 1990, FWS expended approximately $14 million on salmonid/nonsalmonid production and hatcheries (approximately $17.5 million in 1991) and reared more than 75 million anadromous salmonids and nearly 10 million striped bass. In addition, the 1989–1991 research program of FWS allocated annually approximately $2.4 million to studies with direct implications for anadromous or coastal/estuarine species. Nutrition, disease control, and rearing strategies/systems for Pacific salmon, At-

TABLE C-5 U.S. Fish and Wildlife Service Expenditures on Marine Aquaculture, FY 1986–1991 (million U.S. dollars)

	1987	1988	1989	1990	1991
Production/ hatcheries					
Salmonid	10.797	11.137	11.487	12.666	15.618
Nonsalmonid	1.368	1.368	1.409	1.415	1.897
Research					
Salmonid	1.1 (est.)	1.2 (est.)	1.2466	1.3960	1.5125
Nonsalmonid	0.85 (est.)	0.90 (est.)	0.9543	1.2465	0.821
Total	14.115	14.623	15.0969	16.7235	19.8485

SOURCE: Personal communication. John G. Nickum, U.S. Fish and Wildlife Service (1991).

lantic salmon, and striped bass receive the greatest emphasis, but Atlantic sturgeon, shortnose sturgeon, and American shad have been addressed recently. Studies conducted by the National Fisheries Contaminant Research Center and the National Wetlands Research Center have implications for marine aquaculture, but they generally are not focused directly on aquaculture issues. Table C-5 provides an overview of FWS expenditures on marine aquaculture from FY1986–1991.

The role of the FWS in marine aquaculture remains to be concisely defined, especially with respect to estuarine and coastal species. Availability of facilities and staff expertise often determine whether a particular species or problem will be addressed by NMFS or FWS. Cooperative working arrangements unencumbered by tight definitions of traditional agency jurisdiction would be beneficial to the development of marine aquaculture.

NATIONAL SCIENCE FOUNDATION

The NSF Small Business Innovation Research Program provides funding for aquaculture research. This program, open to small business firms, annually solicits high-quality research proposals on important scientific or engineering problems that could lead to significant public benefit. Aquaculture proposals submitted under this solicitation most generally fall under two major suggested topic areas: marine/estuarine or freshwater aquaculture. Certain topic areas in engineering may be appropriate as well, depending on the nature of the proposed research. As Table C-6 shows, the SBIR pro-

TABLE C-6 Proposals and Awards for Aquaculture Research Through NSF Small Business Innovation Research Program, FY 1982–1991 (thousand U.S. dollars)

	1982	1983	1984	1985	1986	1987	1988	1989	1990	1991	Total
Phase I Proposals	27	27	16	26	33	32	17	16	16	9	219
Phase I Awards	4	5	3	2	5	4	3	2	3	1	32
Phase II Proposals	0	2	5	1	0	5	2	5	1	3	21
Phase II Awards	0	1	3	1	0	2	1	3	1	NA	12
Total	119	297	295	615	218	405	394	824	203	50	3,420

SOURCE: Joan R. Mitchell, Ocean Sciences Division, NSF. Personal communication. (1991).

NA: Data not available.

gram has awarded a total of approximately $3.4 million for aquaculture research since FY1982.

REFERENCES

McVey, J. The National Sea Grant College Program Annual Report FY1989, Aquaculture. U.S. Department of Commerce, Washington, D.C.

Regional Aquaculture Centers. 1990. Draft report prepared by the Office of Aquaculture, Cooperative State Research Service. U.S. Department of Agriculture. August.

Appendix D

Sociocultural Aspects of Domestic Marine Aquaculture[1]

SHIRLEY J. FISKE[2] AND JEAN-PIERRE PLÉ[3]

INTRODUCTION

The emergence of a new industry means social change as well as technological change. There is no doubt, for instance, that the domestication of plants for agriculture and the development of animal husbandry vastly changed humans' way of living for millennia. On a smaller and more contemporary scale, the introduction of technologies such as containerization in the maritime industry changed the labor requirements and social organization for servicing vessels; the introduction of purse seines in the tuna fishery initiated distant water fishing, changed labor requirements, and affected fishermen's community and family life. The development of marine aquaculture is no exception.

The introduction of marine aquaculture is not simply the manipulation and culturing of fish and shellfish for human consumption—it will reorganize the social organization for producing fish for consumption, and it involves changing or capitalizing on long-standing cultural attitudes and practices.

In assessing the impact of new industries on social structures, the social sciences look not just at technical factors such as number of ponds, species cultured, or value of product, but at sociocultural aspects of the industry such as the types of units of production, requirements for capital and labor, and distribution of employment opportunities. Social scientists ask who benefits and how are the benefits distributed throughout society? What kind of effect will the technical change have on our social system? Will such a change affect the way of life in communities where marine aquaculture is adopted? How does the introduction of marine aquaculture affect concentrations of capital, and to whom (or where) do the profits flow? Who is employed and at what levels in the system? Is there social mobility between strata in the system? What are the social attitudes or constraints that affect the development of marine aquaculture, and how can these be overcome?

Given the fact that marine aquaculture is the "domestication" of fish and shellfish and the introduction of technology and social structures to accomplish this, the aim of the social sciences is to take a long-term perspective of the associated social costs and benefits of such changes.

SCOPE AND METHODS

The social aspects of the marine aquaculture industry refer to the sociological, demographic, and cultural elements of the marine aquaculture industry. These include the social structure created by the industry itself and how it links with the larger society of which it is a part. Variables considered in an analysis of the social structure include the following:

- the units of production;
- social stratification among owners and between owners and laborers;
- ethnic, demographic, and socioeconomic characteristics of the aforementioned;
- relationship of laborers to production and distribution of capital;
- concentration of producers into vertically integrated firms with processing and marketing or distribution capabilities;
- the capitalization of production (or lack of it);
- the marginalization of coastal laborers or the social mobility of such; and
- the way marine aquaculture relates to the cultural context of a community or region.

Additional areas of interest encompass understanding how the enterprise complements or conflicts with the values that underpin local residents' lives, their work, or their decisions, and how it relates to the established social structure for production of commodities.

Information on social aspects of marine aquaculture in the United States is not readily available, but does exist in agency reports, industry surveys, Sea Grant or U.S. Department of Agriculture (USDA) extension project files. Because of the dispersed and relatively thin nature of sociocultural information, this report is an attempt to develop a conceptual framework for understanding the social aspects of marine aquaculture. Given the paucity of data on the sociocultural aspects of marine aquaculture, the authors conducted an informal telephone survey with Sea Grant extension agents, area specialists, and industry entrepreneurs. A list of respondents is available at the end of the appendix. From these multiple sources, we present several illustrative examples of how marine aquaculture has evolved differently in various parts of the United States.

MARINE AQUACULTURE PROMOTION AS AN EXAMPLE OF PLANNED CHANGE

From a social science perspective, marine aquaculture can be considered as a type of planned change for economic development. Planned change involves the deliberate introduction of new methods of technology or social organization (including values) to change the mode of production of commodities, or behaviors toward any number of economic development goals, including raising fish, shellfish, or mollusks. Currently, the U.S. government encourages the growth of the marine aquaculture industry through funding research and development of new species and new technologies, and through extension efforts to encourage the adoption of such activities, in collaboration with university researchers and private facilities. The implicit goal is to increase the number of producers of farm-reared fish and shellfish. Thus although the United States does not have an explicit marine aquaculture development plan, the effort nonetheless fits the general model of planned economic development.

There are typically three elements in the successful introduction of planned change: (1) appropriate technology; (2) a perceived benefit-flow strategy; and (3) an identified institutional strategy.

Appropriate technology is an integral part of planned change. The majority of the U.S. effort has focused on developing technology (including biotechnology) that will produce a reliable, profitable crop. Appropriate technology also means technology aimed "appropriately" at the adopter, knowledge of the target audience, and awareness of the social and economic effects of adopting the new technology. *Benefit-flow strategy* means that there is a clear flow of benefits and that people can see themselves as part of the process of the flow of benefits. When this occurs, people are more likely to adopt a new way of doing things (like raising fish) or new technology. *Institutional strategy* refers to identifying the best method to implement marine aquaculture through the institutions, nonprofit organizations, academic, and voluntary associations that exist in the target population.

Within the framework of planned change, social sciences can provide information in the following areas: (1) understanding the adoption of marine aquaculture as an enterprise, (2) identifying consumer attitudes and refining strategies for marketing specific products, and (3) identifying unintentional or long-range social impacts.

Understanding Adoption of Marine Aquaculture as an Enterprise

Studies on adoption of innovations are a mainstay of rural sociology and development anthropology. The set of theories for understanding the problems and opportunities in adoption of innovations has been addressed recently by the Farming Systems Research (FSR) approach to rural economic development. The FSR approach has been successful in increasing production of commodities such as genetically manipulated sorghum and millet, according to conditions specified by local farmers. The FSR approach starts with the assumption that farming (or aquaculture) is an integrated and systemic set of activities that includes the household, marketing channels, and production units. Through investigation of the cultural context of the production unit activity, FSR targets the concerns of the household or production unit, such as lack of credit, the lengthy and problematic permitting process, lack of labor to attend to fish ponds or to guard against poachers, or attitudes such as fear of losing freedom and time for other activities. We need to understand the integrated social aspects of farming (or aquaculture) production and perceived constraints in order to encourage adoption. Such a strategy could be fruitfully applied in the case of marine aquaculture development.

Identifying Consumer Attitudes and Refining Marketing Strategies

Social science information can improve acceptance and promotion of marine aquaculture products through understanding consumer attitudes and through marketing. Psychometrics and a branch of anthropology called cognitive anthropology have a long history in studying marketing behavior. A recent example is a southeast Atlantic campaign to promote underutilized species of recreational fish on the basis of studies of fishermen's perceptions and the development of a targeted "marketing campaign" for selected species with a high preference factor.

Techniques similar to this market development can be applied to the introduction of new, hybrid, or transgenic farm-raised animals. Consumer beliefs and attitudes about cultured products have not yet been systematically examined for opportunities to market products. Social sciences are useful in gauging consumer attitudes toward products such as hybrid striped bass or triploid oysters, and in designing marketing campaigns for new products or underutilized products. Each species will have its own set of advantages and disadvantages and can be marketed for growth in niche or mass markets. For example, cultured oysters can capitalize on the health concerns of consumers by promoting the relative bacterial safety of cultured or depurated oysters. Hybrid striped bass has a bland, white meat that may be appropriate for mass rather than niche marketing techniques. It is useful to know the depth, breadth, and dimensions of consumer perceptions, regardless of whether they are judged rational by producers and scientists. Growing consumer concerns, such as the belief that antibiotics are overused in pen-reared fish or the desire by consumers to avoid wild-caught fish because of by-catch issues, are potential candidates for promotional campaigns, but it is necessary first to understand the dimensions of these consumer perceptions.

Identifying Unintentional and Long-Range Social Impacts

It is important to anticipate the challenges and opportunities marine aquaculture is likely to face, and to know the direction in which the industry is headed. In terms of community acceptance, social sciences are useful in anticipating community resistance on issues of multiple use, wastewater concerns, siting, and aesthetics, or the value of marketing marine aquaculture as a solution to problems regarding endangered species by-catch or overfishing of wild stock.

Observations from outside the United States on social aspects of marine aquaculture in developing countries can provide valuable insights into these issues (Meltzoff and LiPuma, 1985–1986; Bailey, 1988; Shang, 1990; Weeks, 1990). While the context of marine aquaculture development is very different abroad, the underlying processes and issues are comparable to the situation in the United States. The investigators have noted that the nutritional base of local populations suffers as the production shifts toward the lucrative export market; that the development of marine aquaculture affects land values, resulting in the displacement of certain groups; that the water body becomes privately controlled as opposed to open to multiple use and public access; that the local employment base is augmented; that there is not much mobility among labor, manager/technical, and ownership roles; and that changes in production techniques and processes of economic development affect the amount and distribution of community wealth (Smith, 1991).

The focus of marine aquaculture development has implications for the mix and provision of services such as extension, education, credit, and research and development. For example, if the objective is to encourage investment in marine aquaculture by small producers such as household units, the extension effort would be different than if the aim is to encourage individuals who have access to venture or foreign capital.

To carry the example further, if the goal is to create a labor-intensive industry that contributes income across a broad population base, it seems logical to concentrate on those types of marine aquaculture that promote that goal (e.g., soft-shell crabs or small-scale pen rearing of salmon as in Maine). On the other hand, if the goal is to produce large amounts of biomass as efficiently as possible with intensive aquaculture methods (where potential aquaculturists are likely to be experienced businessmen and women diversifying from other industries such as liquor distilling or poultry), the potential adopters likely have different research or extension needs than household production units.

Finally, demographic information on the developing United States marine aquaculture industry helps forecast labor, education, and training requirements. Social scientists can help identify the challenges that lie ahead and must be faced—including contentious issues of conflicts with recreational boating, multiple use, and questions about who benefits (distributional questions) and about who bears the costs (externalities). Anticipating these areas of potential conflict ahead of time helps avoid costly court battles, lawsuits, and social protest. It promotes effective advocacy for an emerging industry that is realistically based on facts and knowledge about the social milieu.

Up to now, we have focused on the conceptual framework of marine aquaculture as a type of planned change and the value of addressing sociocultural aspects of marine aquaculture. To make this more concrete, we turn now to selected illustrative examples in marine aquaculture development in order to highlight the differences and variability in social and cultural aspects.

ILLUSTRATIVE EXAMPLES

The purpose of this section is to provide snapshots of various forms of marine aquaculture in the United States. These examples show that the forms that marine aquaculture takes depend on the historical and cultural background of specific regions in the United States, legislative requirements in individual states, the capital requirements of particular species, land use and demographic constraints in particular areas, and many other factors. We have chosen a few marine aquaculture efforts that are already technically and economically feasible and are fairly well established. Given the paucity of systematic data, these cases serve as illustrative examples of the diversity in social structure and sociocultural dimensions involved in marine aquaculture in the United States.

Oyster Growing in the State of Washington

The state of Washington is one of the leading producers (by volume) of cultured oysters. The social organization is traditional and is characterized by multigenerational enterprises organized around kin-based lineages. Several (four to five) of these family operations operate vertically integrated enterprises from the hatchery to processing operations.

The origins of the social organization of this industry trace back to the nineteenth century when Washington joined the Union and the new state legislature passed a law permitting public tidelands to be deeded to private individuals for the express purpose of oyster farming. The law stipulated that such deeded lands must remain in continuous oyster production or ownership would revert back to the state. As a result, many of the oyster farms currently in operation are well-established commercial enterprises in their second or third generation of continuous ownership. Deeded areas can be transferred between parties at market prices, with the condition that the new owner must continue to use the tidelands to farm oysters. Several of these deeded farms also operate their own hatcheries and process their oyster harvest.

The state no longer sells deeds to submerged lands, and a moratorium on leasing public tidelands was enacted in the late 1980s. No public funds are expended to seed deeded farms, but the state does support a program to seed public oyster beds. The newer farms on leased tidelands tend to be smaller enterprises than the original family farms on deeded lands.

The oyster farms in Washington generally tend to be labor intensive, not necessarily in terms of numbers of laborers required but in terms of functions performed. This is especially true of harvesting and shucking oysters. Seasonality of labor supply is not an issue. Oysters can be raised and harvested year-round.

Labor requirements are a function of the size of particular firms and the management skills of individual operators. Although Caucasians are the dominant ethnic group for operators and laborers, many Asians are now entering the industry as laborers. Those entering the industry as owners/ operators typically have some background in oyster raising since there is long tradition of the industry in the state.

The primary difficulty in expanding oyster production in Washington does not appear to be a lack of capital because large amounts of money are not needed to start raising oysters. Instead, the constraints are physical—lack of suitable locations. The best sites are already producing oysters, and water quality problems and multiple-use conflicts render other sites less desirable for raising oysters (even if the leasing moratorium were to be lifted).

Salmon Ranching in Alaska

Alaska's cold and clean waters, large expanses of uninhabited areas, and protected waterways are all desirable attributes for siting fish farms. Amidst the opportunity for lucrative private enterprise, the state of Alaska has tried to balance public good and private interests through a unique institutional arrangement for salmon aquaculture. There is a prohibition on private, for-profit salmon farming (pen rearing and sale of fish without release into the common property fishery), but salmon ranching (the release of hatchery-raised fish to supplement wild stocks) is permitted and is carried out by a mix of state-owned hatcheries and licensed, private, nonprofit (PNP) hatcheries. Currently, there are 19 hatcheries operated by the state and 21 operational PNP hatcheries.

Salmon ranching was authorized by the state legislature in the early 1970s in response to an ailing salmon fishing industry. Catches were at all-time lows, and the development of an enhancement program to serve the state in years of both lean and bountiful salmon harvests was envisioned by Alaskans. The state hatchery program was initiated in 1971 with the creation of the Division of Fisheries Rehabilitation, Enhancement, and Development (FRED) in the Alaska Department of Fish and Game. In 1974, the state-owned and managed program was expanded to permit private nonprofit salmon ranching. The intent was to authorize the private ownership of salmon hatcheries by qualified nonprofit corporations.

Permits for PNP salmon ranching are granted by the state to qualified private individuals or associations. The permits are nontransferable and remain valid until rescinded by the state. Since 1974, only three PNP salmon ranching facilities have failed, which suggests that private nonprofit salmon aquaculture can be highly successful. The incentive to become involved in nonprofit salmon ranching is twofold. First, operators are permitted to use revenues to pay for annual operating loans, capital loans, and staff salaries. Second, the large PNP corporations have a Board of Directors whose members may include representatives from environmental, scientific, recreational, and subsistence harvest interests, but it consists primarily of commercial fishermen representing trollers, seiners, and gill-netters. These fishermen stand to profit from the release of millions of additional juvenile salmon (925.2 million in 1990) that can be harvested later during commercial openings in the various fisheries.

The PNP hatcheries include both large operators (about one-fourth showed revenues of more than $1 million) and smaller operators, commonly described as the "mom-and-pop" variety. The larger salmon hatcheries tend to be operated by regional aquaculture associations composed of limited entry license holders (the commercial fishermen noted above), and they usually employ up to 30 individuals on a full-time basis, with additional employees on a seasonal basis. The smaller operators consist largely of individuals with backgrounds in the fishing industry, and employ approximately five individuals on a full-time basis and hire additional employees for seasonal work.

The concept of private for-profit salmon ranching is not likely to be considered in the state of Alaska, particularly as long as wild-harvest fishermen dominate the politics of state fisheries. Over the last three years, however, operational responsibilities for some of the state-operated hatcheries have been transferred to PNP regional aquaculture associations; some state policymakers, including some members of the Alaska Legislature, believe that the state should no longer be in the business of operating those public hatcheries that benefit primarily commercial fishermen.

Baitfish, Trout, and Yellow Perch in the Great Lakes

Even though they represent freshwater systems, it is important to address aquaculture in the Great Lakes region for several reasons. Despite the fact that these states are not "ocean" states, they are "marine" states in that they are eligible to participate in many federal marine programs such as coastal zone management, marine sanctuaries, Sea Grant, and national seashores. Additionally, the Great Lakes states are coastal states, and aquaculture development in this region may experience the same conflicts and challenges as the conventional coastal states have experienced, particularly with respect to sociocultural aspects. Developments in Minnesota and Michigan are addressed next.

Minnesota

Aquaculture in Minnesota can be traced back to nineteenth century European immigrants who practiced trout farming in Scandinavia. Today, there are about 150 licensed aquaculture producers in Minnesota, of which 75 are involved in baitfish species, 50 in trout farming, and the reminder in other species (walleye, sunfish, paddlefish, and carp) used for private stocking purposes. Baitfish aquaculture is supported by an additional 475 licenses for the wild harvest of baitfish. Holders of wild-harvest baitfish licenses supply baitfish aquaculture operators with wild stock for grow-out. Baitfish aquaculture has come to dominate fish farming activities in Minnesota through an elaborate intra- and interstate distribution network, resulting in an industry now worth $50 million annually. In contrast to baitfish, nearly all of the farm-raised trout is consumed locally and the industry value is considerably smaller.

It appears that three main social groups have engaged in aquaculture in Minnesota. The first group includes individuals traditionally associated with harvesting wild resources through trapping, hunting, and fishing. This group specializes largely in the baitfish sector, both as wild harvest collectors and as aquaculture operators. The second group includes farmers and other landowners who have a water resource (i.e., a pond or a lake) on their land and see aquaculture as a supplemental source of income. The third group involved in aquaculture is entrepreneurs with engineering or science backgrounds.

Michigan

As in Minnesota, aquaculture in Michigan began in the nineteenth century, and some of the fish farms in operation today began during the 1920s. In contrast to the situation in Minnesota, Michigan aquaculture is dominated by trout farming—rainbow and brook. Other farmed species include large- and smallmouth bass, sunfish, yellow perch, walleye, and baitfish. Of the 110 aquaculture license holders in the state, only about 20 percent could be considered profit-motivated business operations, with the remainder involved in aquaculture for various other reasons. Of these operators, five to six farms produce 80 percent of the trout. These farms employ seven to eight individuals year-round, with another one to two individuals hired during harvest periods. The majority of operators involved in aquaculture are mom-and-pop types of operations or small producers primarily interested in earning supplemental income. Some of these smaller operators are also involved in related businesses such as trout brokering and programs promoting aquaculture at county fairs and other events.

The state of Michigan has begun leasing older state hatchery facilities, which have been retired from state service to local governments. The hatcheries are then leased to private aquaculture operators who operate the hatcheries as profit-making fish farms and in some cases as local tourist attractions. The leasing of the state's hatchery facilities to local governments is an example of an institutional strategy designed to promote and maintain a public role in Michigan's aquaculture development.

Aquaculture can serve to support and maintain regional cultural traditions. For instance, in many parts of the Upper Midwest (particularly Wisconsin and Michigan), the "Friday night fish fry" is a local cultural tradition. The preferred fish for fish fries is yellow perch, but Great Lakes yellow perch harvests have declined significantly. The preference for yellow perch is so strong that area restaurants are willing to pay much higher prices (sometimes in the neighborhood of four to five times) for yellow perch when it is available, in lieu of the other kinds of fish. Currently, technology constraints prevent the development of a yellow perch aquaculture industry. If such constraints could be overcome, yellow perch aquaculture could reduce stress on wild stocks, reduce the region's dependence on outside sources for fish, and help promote and continue a local cultural tradition.

Texas Shrimp Farming

With imports contributing nearly 75 percent of the shrimp consumed in the United States, there would seem to be a strong incentive to develop a viable domestic shrimp marine aquaculture industry in this country. However, along the Gulf Coast from Texas to Florida, the part of the country that is most suitable for shrimp aquaculture, shrimp farms operate only in the state of Texas.

The social structure of shrimp farming in Texas is characterized by two opposite types: large and heavily capitalized intensive shrimp production with foreign financing or venture capital, and small producer firms, primarily family operations in which shrimping is one of a number of income-generating strategies, fitting the model of "occupational multiplicity."

Large-scale intensive shrimp aquaculture in Texas is currently performed by two large commercial operators, both owned by Taiwanese interests. The smaller-scale operations are comprised of about six to eight smaller producers, generally family-operated enterprises. In this latter category, only one operator is considered a commercial enterprise (i.e., a for-profit business). The remaining small operators engage in marine aquaculture to supplement their income as one of a number of income-generating strategies. Such occupational multiplicity is an important strategy for ensuring family income during lean times.

People entering shrimp marine aquaculture at the small producer level include shrimpers who have left the fishery and individuals with backgrounds in poultry, among others. They tend to be of Caucasian origin, while individuals of other ethnic backgrounds are employed as staff.

Labor requirements per acre are small, with a slight increase necessary during harvest time or if operating a hatchery. Workers usually acquire their skills from on-the-job training, and individuals with advanced degrees are increasingly available. Many of the marine aquaculture specialists coming out of Texas universities are finding jobs in Latin America where there are more employment opportunities than in Texas.

Difficulties in increasing production include regulatory and financial constraints. The state recently banned the importation of exotic species of fish, including juvenile shrimp from out-of-state hatcheries. The only in-state hatchery, operated by one of the large Taiwanese farms, is not able to supply all the needs of the other Texas farms. As a result, many of the small producers are not able to stock their ponds and consequently are not able to produce a harvest. The financial constraint is attributable to the recent condition of the Texas banking sectors in which many banks and savings and loan companies have been forced into bankruptcy. A potential investor in marine shrimp aquaculture in Texas would have to depend almost exclusively on private investment capital.

The political and sociocultural milieu for marine aquaculture in Texas is improving. There has been a recent shift in state responsibility for marine aquaculture development from the Department of Natural Resources to the Department of Agriculture, and appointment is expected of a state marine aquaculture liaison officer to coordinate development between the state agencies and the legislature. Traditional Texan cultural values also favor marine aquaculture development. Texans place high esteem on independent farming activities. Small-scale shrimp farms have typically been welcomed, and there does not seem to be strong adverse reaction to the ownership and development of the two largest shrimp farms in Texas by Taiwanese interests. Commercial shrimpers apparently accept shrimp marine aquaculture operations. The lack of opposition may be attributable to the small total output of marine shrimp aquaculture, the fact that shrimp aquaculture does not interfere with marine navigation, or the view that shrimp aquaculture is a means to reduce the importation of shrimp.

Soft-Shell Crabs in Mid- and South Atlantic

The expansion of the soft-shelled crab industry is a success story that is intimately tied with the social fabric and culture of commercial watermen in the mid- and south Atlantic. The 1988 gross value of the crop has been estimated to be approximately $4.5 million. The development of soft-shell crab marine aquaculture has benefited small producers, in this case families, in particular, families who have been traditionally engaged in seafood harvesting as crabbers or watermen. Most soft-shell crabs are collected by watermen and monitored by their families using simple, often homemade methods such as holding pens. Some use more sophisticated recirculating flow-through systems. There is usually little value-added work to the product by the producer, and most crabbers are not involved in distributing or marketing the crabs.

Sea Grant marine extension specialists and agents have worked with crabbers to improve the holding and handling techniques, and the effort has paid off. Total labor involved in harvesting and supervising the crabs throughout the Southeast and mid-Atlantic probably exceeds that of other types of producers, and a decentralized approach to supply and production has evolved to date. Small-scale fisheries tend to provide greater employment opportunities than large-scale enterprises. Additionally, small-scale fisheries often allow opportunities for laborers to become owners of the production unit, whereas large heavily capitalized farms often preclude that possibility (Pollnac and Poggie, 1991). In the United States, there is an unspoken judgment that capital intensive is better because it is likely to use capital more efficiently. The social impact, however, is the trade-off between greater employment opportunities for people and the potential for social mobility in small-scale marine aquaculture versus the efficiency of more highly intensive and capitalized marine aquaculture operations.

In soft-shell crab production there is usually a division of labor within the household, with wives in charge of monitoring the progress of the peelers. This takes close supervision and requires checking the pens every few hours. Although biologists typically see the constraints on this industry as the availability of a supply of peelers, other constraints include the availability of family labor to monitor the molting crabs because the "window of opportunity" to harvest them is narrow.

Soft-shell aquaculture development in Georgia is typical of the situation throughout the mid-Atlantic states, although the numbers of families or individuals involved is much greater in Maryland, Virginia, and North Carolina than in Georgia. There are around 14 to 15 soft-shelled crab producers in the state of Georgia, employing about 30 individuals. All but one are mom-and-pop operations run by active crabbers. The men usually harvest and/or purchase peeler crabs from other crabbers, and the women help collect the shed crabs from facilities adjacent to the home. All the operators are Caucasian, although one wife is Asian. Most soft-shell crabs are shipped frozen or live to northern markets with little further processing. This fits the model of the household as the production unit, division of labor by age and sex, and product or occupational multiplicity within the family-based production unit.

The growth in popularity and production of soft-shell crabs has fueled the development of specialized processors. At least one soft-shelled crab entrepreneur on Maryland's Eastern Shore has taken advantage of international demand for soft-shelled crabs. The businessman, who was an executive in the poultry industry, has a sophisticated processing and packaging plant that relies on local watermen to collect and deliver crabs to his processing plant. From the processing plant the soft-shelled crabs are exported directly to Japan.

Salmon and Mussels in Maine

To the people of Maine, particularly those living in remote coastal communities, the ocean is traditionally viewed as a working resource that provides economic opportunity. With these values as background, marine aquaculture has been largely accepted on its merits, namely, as another useful way to obtain products from the sea. Early in the development of aquaculture, there was opposition because of fears that the ocean bottom would be "rented" exclusively by a few individuals. Today marine aquaculture is viewed less as a competitor and more as a natural expansion of the state's fishery resources and as an opportunity to diversify sources of income. State law upholds the importance of protecting traditional fishing rights and giving priority consideration to the needs of lobstermen and shrimpmen by prohibiting marine aquaculture development in areas that interfere with traditional lobster grounds or with established nearshore navigation routes.

In the process, marine aquaculture in Maine has been transformed from a white-collar occupation pursued by wealthy outsiders, investors, scientists, and the curious, into a blue-collar occupation in the same tradition as fishing for herring, scallops, or lobsters. This transformation occurred as the potential for profits from marine aquaculture became evident.

Pen rearing of salmon is the most successful type of marine aquaculture practiced in Maine, with a projected 1992 value of $88 million (by comparison, lobster was valued at $46 million in 1989). There are four or five large salmon producers, but the majority of salmon farms (about one-half to two-thirds) are small producers—family operations employing three to seven workers. Individuals generally come from fishing backgrounds, such as lobster, herring, or scallops, or else from other marine industries. The individuals are most often Caucasian and could be characterized as "Down Easters"—families that have lived for generations in eastern Maine.

Marine aquaculture developed in Maine under the politically expedient premise that it could not interfere with the operations of traditional fisheries. As a result, salmon pens have had to be located in remote bays and coves away from traditional fisheries, particularly away from inshore areas used by lobstermen. This requirement not only satisfied the needs of wild-harvest fishermen, but in turn provided a new economic resource to those isolated communities in which the pens were located. Recent research suggests that aquaculture operations have enhanced lobster harvest in areas under the fish pens. Recognizing these mutual benefits, a lobster cooperative on Swans Island has entered into a formal agreement with salmon farm operators permitting the salmon farmers to tap into lobster market and distribution channels.

The culture of mussels has also become a recognized addition to Maine's working ocean concept. The state currently has one large mussel producer employing 35 full-time and 30 part-time workers, but many more mussel operations are small, family-type producers with two to three workers each. All mussel farms are located on sites leased by the state. For these small producers, mussel culture is usually an on-the-side enterprise in addition to their regular occupations as lobstermen, oystermen, or shrimpmen. As in the case with salmon, mussel culture is viewed not as a competitor but rather as a legitimate partner among all ocean-related industries.

CONCLUSION

In addition to a change in technology from capturing wild fish to raising them in captivity, the introduction of marine aquaculture entails social changes. Domestic marine aquaculture development is likely to cluster around two general social structural types. The first type is small-scale aquaculture where the household or extended family is the production unit (such as soft-shell crabs, small-scale salmon pen rearing, some shrimp aquaculture). Aquacultural production will probably be one of a number of occupations being exploited at any one time (occupational multiplicity). This type of social arrangement for aquaculture preserves the relationship of producer to the product and generates income across large numbers of households. It helps ensure economic stability in rural coastal areas that often are economically depressed. The social value of marine aquaculture to a broadly distributed population of family-based units of production, in which aquaculture is predominately a part-time occupation, should not be overlooked or underestimated. Income distribution is an important aspect of economic and family stability in rural and economically depressed areas.

A second type of marine aquaculture seems to be heavily capitalized enterprises, backed by industry, venture, or foreign capital. These businesses are likely to be intensive shrimp farms or large-scale salmon pen rearing operations. In terms of producing greater amounts of fish for consumption by the American public and offsetting U.S. balance of payments deficits, this type of aquaculture development is valuable. In short, there are important roles for each type of aquaculture.

Undoubtedly, many questions remain including what role marketing can play in improving marine aquaculture acceptance by the public; who the potential beneficiaries are among social groups to further marine aquaculture development; and what sociopolitical conflicts will have to be resolved to maximize marine aquaculture growth. Answers to these questions and further regulatory and technological advancements can help the United States maximize its marine aquaculture potential.

ACKNOWLEDGMENTS

The authors wish to acknowledge the cooperation of the following individuals for providing background information or assistance in preparing the vignettes included in this report.[4] Dallas E. Alston, Department of Marine Sciences, University of Puerto Rico; Bob Brick, Aquaculture Management Associates, College Station, Texas; Fred S. Conte, Aquaculture Specialists, Cooperative Extension, University of California, Davis; Rick DeVoe, South Carolina Sea Grant Consortium; David Dow, Executive Director, Lobster Institute and Sea Grant Extension Specialists, Maine Sea Grant Program; Kevin Duffy, Salmon Rehabilitation and Enhancement Coordinator, Alaska Department of Fish and Game; Mike Ednoff, Aquaculture Development Representative, Florida Department of Agriculture and Consumer Services; John Ewart, Delaware Sea Grant Marine Advisory Service; Don Garling, Professor, Department of Fisheries and Wildlife, Michigan State University and Extension Fisheries Specialists; David Landkamer, Assistant Aquaculture Extension Specialists, Minnesota Department of Fish and Wildlife; Carter Newell, Great Eastern Mussel Farm; Terry Nosho, Aquaculture Extension

Specialist, University of Washington Sea Grant; Mac V. Rawson, Director, University of Georgia Sea Grant College Program.

NOTES

1. The views expressed herein represent the authors' and do not necessarily represent the views of the federal government, the National Oceanic and Atmospheric Administration, or the National Sea Grant College Program.

2. Program director for Social Science and Marine Policy, National Sea Grant College Program.

3. Doctoral student, College of Marine Studies, the University of Delaware.

4. The views expressed by these individuals do not represent the official position of any organization with whom they may be affiliated.

REFERENCES

Bailey, C. 1988. The social consequences of tropical shrimp mariculture development. Ocean and Shoreline Management 11: 31-44.

Meltzoff, S., and E. LiPuma. 1985–1986. The social economy of Coastal resources: Shrimp Mariculture in Ecuador. Culture and Agriculture No. 28 (Winter).

Pollnac, R.B., and J.J. Poggie. 1991. Introduction. Pp. 1-18 in Small-Scale Fishery Development: Sociocultural Perspectives, J. Poggie and R. Pollnac, eds. Kingston, R.I.: International Center for Marine Research Development, University of Rhode Island.

Shang, Y.C. 1990. Socioeconomic constraints of marine aquaculture in Asia. World Marine Aquaculture 21(1):34-43.

Smith, C.L. 1991. Patterns of wealth concentration. Human Organization 50(1):50-60.

Weeks, P. 1990. Marine aquaculture development: An anthropological perspective. World Marine Aquaculture 21(3):69-74.

Appendix E

Committee Member Biographies

ROBERT B. FRIDLEY (NAE) received an A.A. from Sierra College; a B.S. (mechanical engineering) from the University of California, Berkeley; an M.S. (agricultural engineering) from the University of California, Davis; and a Ph.D. (agricultural engineering) from Michigan State University. Dr. Fridley is currently Executive Associate Dean of the College of Agricultural and Environmental Sciences at the University of California, Davis, where he previously served as director and professor of the Aquaculture and Fisheries Program. Dr. Fridley also served as chairman of the Agricultural Engineering Department at Davis. He formerly worked for Weyerhaeuser Company where he managed various silvicultural and horticultural, as well as aquacultural, engineering programs. Dr. Fridley has conducted research and been responsible for research management and academic administration related to engineering, design and development of production systems in aquaculture, fisheries, horticulture, and silviculture; and engineering education. He is active in the technical committees of the American Society of Agricultural Engineers, the American Society for Engineering Education, the American Fisheries Society, and the World Aquaculture Society.

JAMES L. ANDERSON received a B.S. (biology/economics) from the College of William and Mary, an M.S. (agricultural economics) from the University of Arizona, and a Ph.D. (agricultural economics) from the University of California, Davis. He is currently an assistant professor of economics at the University of Rhode Island. Dr. Anderson is involved in extensive teaching and research in aquacultural economics, economics of seafood quality, and seafood marketing and trade. He has served as advisor to the Canadian Department of Fisheries and Oceans, the Alaska Department of Commerce and Economic Development, and several private firms. Dr. Anderson is an associate editor of *Transactions of the American Fisheries Society* and is on the Editorial Council of the *Journal of Environmental Economics and Management.* He has participated as a member of the Technical Advisory Council to the Northeast Regional Aquaculture Center, has served on the planning committee for the National Research Council (NRC) study on the Assessment of Technology and Opportunities for Marine Aquaculture in the United States, and is a member of numerous professional economic, fisheries, and aquaculture societies.

JORDON N. BRADFORD has received technical training in industrial turbines, natural gas engines, blueprint analysis, water pumps, magnetos, solid-state ignitions, and pneumatics. He is currently president of Oyster Farms, Inc., of Louisiana; owner/operator of Bradford Oyster Company, Inc., of Mississippi; and vice-president of Gulf Shellfish of Texas. He has been involved in the development and establishment of the first commercial shellfish hatchery in the Gulf of Mexico region and the first transfer of state-of-the-art remote setting technology to the Gulf region.

BILIANA CICIN-SAIN received the A.B., M.A., and Ph.D. degrees in political science from the University of California, Los Angeles, and postdoctoral training at Harvard University. She is currently professor in the Graduate College of Marine Studies, University of Delaware and codirector of the university's Center for the Study of Marine Policy. She previously served as acting director of the Ocean and Coastal Policy Center, Marine Science Institute (of which she was a founder), and associate professor of political science at the University of California, Santa Barbara. Dr. Cicin-Sain has been a guest scholar or fellow at the Environment and Policy Institute (summer 1988); Brookings Institution (fall 1984); Rockefeller Foundation and Study Center (October 1984); Woods Hole Oceanographic Institution (1983–1984); and the East-West Center (1988). Dr. Cicin-Sain is well known for her work on national, international, state, and local issues dealing with the institutional framework associated with multiple-use conflicts in the coastal zone. Among her current activities, she is editor of the international journal *Ocean and Shoreline Management*, cochair of the Marine Affairs and Policy Association, and a member of the Ocean Studies Board of the National Research Council.

PAMELA K. HARDT-ENGLISH has more than 10 years practical and research experience in private industry and experience in the areas of food science and agricultural engineering. She received a B.A. (computer science) from the University of California, Berkeley, an M.S. (agricultural engineering) and an M.S. (food science) from the University of California, Davis. She is currently president of Pharmaceuticals and Food Specialists in San Jose, California. Ms. Hardt-English has been instrumental in bringing innovative solutions to bear on transporting, processing, and packaging of agricultural products and associated quality control concerns. Her areas of expertise include process development and verification, evaluating process problems and deviations, automated batch systems, developing quality assurance monitoring, and improving production efficiency for grain, fruit, vegetables, meat, fish, and milk products.

G. JOAN HOLT has more than 18 years experience with warm water finfish research. She received a B.S. (biology) and an M.A. (biology) from the University of Texas at Arlington, and a Ph.D. (fisheries science) from Texas A&M University. She is currently a research scientist at the Marine Science Institute, Port Aransas, for the University of Texas at Austin. She is nationally and internationally recognized for her research in physiology, recruitment, and rearing of larval marine finfish. She was instrumental in developing the technology and understanding to successfully grow red drum from eggs through adults. Dr. Holt is a member of numerous professional associations and has been an active committee and panel participant in the American Fisheries Society (president, Early Life History Section, 1986–1988), organizer and chairman of Ninth Annual Larval Fish Conference (1985) and the Symposium on Larval Fish Recruitment (1986), and member of the organizing committee for the Red Drum Conference (1987).

RONALD D. MAYO received a B.S. in civil engineering from the University of Washington, where he continued graduate studies in sanitary and hydraulic engineering. He is a registered civil and sanitary engineer in the state of Washington. Mr. Mayo is currently principal engineer and director of fish culture projects for James M. Montgomery, Consulting Engineers, Inc., throughout the United States and internationally. He was formerly owner and principal of The Mayo Associates (TMA) in charge of fisheries and fish culture activities. Prior to the formation of TMA, Mr. Mayo was executive vice-president of Kramer, Chin and Mayo and manager of their hatchery and aquarium projects. He is recognized internationally for the planning and design of fish hatcheries and private aquaculture facilities, the development of bioengineering technology, and the planning and evaluation of hatchery systems. One of his most recent projects was an assessment of private salmon ranching in Oregon for the Oregon Coastal Zone Management Association, Inc. Currently, he is involved in planning fish propagation facilities in Washington, Alaska, Oregon, California, New Mexico, Massachusetts, and Ecuador.

KENNETH J. ROBERTS received the B.S. and M.S. degrees (agriculture economics) from Louisiana State University and earned a Ph.D. in resource economics from Oregon State University. He currently has joint appointments as professor in the Center for Wetland Resources of the Louisiana State University, and Marine Economist Specialist for the Louisiana Cooperative Extension Service. He has been active in the economic elements of fishery resource use as related to private and public management, production, and marketing of marine food products; recreational demand for public resources; and marine extension education. Dr. Roberts has been active in the state of Louisiana's Seafood Marketing Task Force, the Louisiana Shrimp Management Task Force, and the Economics Section of the American Fisheries Society (president 1984–1985). He has served on the editorial board of the *North American Journal of Fisheries Management* and is currently economics associate editor for the World Aquaculture Society.

JOHN H. RYTHER received the A.B. (cum laude), M.A., and Ph.D. (biology) degrees from Harvard University. He also received an honorary D.Sc. from the University of New Hampshire in 1979. He is currently a scientist emeritus at Woods Hole Oceanographic Institution, where he was formerly director of the Coastal Research Center, chairman of the Department of Biology, and senior scientist. He is also adjunct professor of aquaculture for the University of Florida, and a former director of the Division of Applied Biology for the Harbor Branch Oceanographic Institution in Florida. Dr. Ryther has served on past NRC committees and commissions, including Committees on Oceanography, Ocean Engineering, Coastal Waste Management, and Aquaculture, and the Commission on Natural Resources.

PAUL A. SANDIFER received a B.S. degree (biology) from the College of Charleston and a Ph.D. (marine science) from the University of Virginia. He currently is director of the Division of Marine Resources, South Carolina Wildlife and Marine Resources Department, and of the James M. Waddell, Jr. Mariculture Research and Development Center. In addition, Dr. Sandifer serves on the faculties of Clemson University, the University of South Carolina, the Medical University of South Carolina, and the College of Charleston. These activities involve frequent interaction with marine resource users, legislators, university faculty and administrators, state and federal agencies, and the general public. Dr. Sandifer is a member of numerous professional societies and has been active in the technical committees of the American Fisheries Society, American Association for the Advancement of Science, National Shellfisheries Association, Caribbean Aquaculture Association, and World Aquaculture Society. He authored the Freshwater Prawn Species Plan portion of the National Aquaculture Plan.

EVELYN S. SAWYER received a B.A. (biology) from the University of Maine, and an M.S. and Ph.D. (zoology) from the University of New Hampshire. She has 18 years of research and industry experience with salmon culture in the northeastern United States. Dr. Sawyer has been president and chief executive officer of Sea Run Holdings, Inc., since 1978. Previously Dr. Sawyer was biologist and partner in Marine Salmon Farms, Inc., and a research faculty member at the University of New Hampshire. She has been active in numerous professional organizations and commissions: the World Aquaculture Society; American Fisheries Society, Treasurer, Fish Culture Section (1987–1988); Maine Aquaculture Association Board Member (1984–1988) and vice-president (1988); and New England Aquaculture Center Technical Advisory Committee (1987–1990).

Appendix F

Participants in Special Sessions

Invited Guests, Committee Meeting in Davis, California, March 1990

Roger Garrett (organizer), Aquaculture and Fisheries Program, University of California at Davis
Harvey Collins, Sacramento Department of Health Services
Fred Conte, University of California, Aquaculture Extension
John Forster, Sea Farms of Norway
Robert Meek, Ecomar, Inc.
Michael Neushul, University of California Marine Science Institute
James Rote, California Joint Committee on Fisheries and Aquaculture
Peter Scrivani, Pacific Mariculture, Inc.

World Aquaculture Society Meeting, Halifax, Nova Scotia, June 1990

Kenneth Chew, University of Washington
John Corbin, Hawaii Aquaculture Development Program
Richard DeVoe, South Carolina Sea Grant Consortium
Carol Engle, University of Arkansas, Pine Bluff
Gery Flynn, Brandan Ur Atlantach, Ireland
John Forster, Sea Farms of Norway
Judith Freeman, Washington Department of Fisheries
Richard Gowen, Dunstaffnage Marine Laboratory, Scotland
Upton Hatch, Auburn University

Thomas Losordo, North Carolina State University
John Pitts, Washington Department of Agriculture
Robert Pomeroy, Clemson University
Gary Pruder, Oceanic Institute, Hawaii
Harald Rosenthal, University of Kiel, Institute of Marine Studies, Germany
Jeff Taylor, Zeigler Brothers
Dallas Weaver, Scientific Hatcheries
Alex Wypizinski, Rutgers University

Invited Guests, Committee Meeting in Hilton Head, South Carolina, August 1990

Richard DeVoe, (organizer), South Carolina Sea Grant Consortium
Eddie Gordon, South Carolina Crab Company
Bennett Helms, International Mariculture Resources, Inc., South Carolina
Ronald Hodson, Sea Grant College Program, University of North Carolina
Jurij Homziak, Mississippi Sea Grant Advisory Service
Fred Kern, National Marine Fisheries Service Oxford Laboratory, Maryland
Ronald Malone, Department of Civil Engineering, Louisiana State University
Steve Otwell, Department of Food Science and Human Nutrition, University of Florida
Wynn Pettibone, Laguna Madre Shrimp Farms, Texas
William Rutledge, Texas Parks and Wildlife Department
Jack Whetstone, South Carolina Marine Extension Service

Aquaculture International Congress and Exposition, Vancouver, British Columbia, September 1990

Alan Archibald, British C. Salmon Farmers Association, Canada
Amanda Courtney, British Trout Association Ltd., England
Eric Edwards, Shellfish Association of Great Britain
David Egan, The DPA Group, Canada
Tony Fox, Fanad Sea Fisheries Ltd., Ireland
Erik Hempel, Dar es Salaam, Tanzania
Peter Hjul, AGB Heighway Ltd.
Lee Lippert, Lippert International, United States
Ted Needham, The DPA Group, Canada
David Rackham, Hydro Food Products Ltd., Scotland
Ronald Roberts, University of Stirling, Scotland
Susan Shaw, University of Stirling, Scotland
John Spence, British Columbia Aquaculture Research and Development Council

Index

A

B

C

D

G

H

N

O

P

R

S

T

U